A Focus on Fractions

A Focus on Fractions is the first book to make cognitive research on how students develop their understanding of fraction concepts readily accessible and understandable to pre- and in-service K-8 mathematics educators. This important resource assists teachers in translating research findings into their classroom practice by conveying detailed information about how students develop fraction understandings as well as common student misconceptions, errors, preconceptions, and partial understandings that may interfere with students' learning. Using extensive annotated samples of student work, as well as vignettes characteristic of classroom teachers' experiences, this book equips educators with knowledge and tools to reveal students' thinking so that teachers can modify their teaching to improve student learning of fractions concepts.

Special Features:

- *End of Chapter Questions* provide teachers the opportunity to analyze student thinking and consider instructional strategies for their own students.
- *Instructional Links* help teachers relate concepts from the chapter to their own instructional materials and programs.
- *Big Ideas* and *Research Reviews* frame the chapters and provide a platform for meaningful exploration of the teaching of fractions.
- *Answer Key* posted online offers extensive explanations of in-chapter questions.

A Focus on Fractions bridges the gap between what mathematics education researchers have discovered about the learning of fraction concepts and what teachers need to know to make effective instructional decisions.

Marjorie M. Petit is an educational consultant in mathematics instruction and assessment and has provided leadership in the development of the Vermont Mathematics Partnership Ongoing Assessment Project.

Robert E. Laird works with the Vermont Mathematics Initiative and is a Lecturer in Mathematics at the University of Vermont.

Edwin L. Marsden is Professor of Mathematics and former chair of the Department of Mathematics at Norwich University.

Studies in Mathematical Thinking and Learning
Alan H. Schoenfeld, Series Editor

A Focus on Fractions

Bringing Research to the Classroom

Marjorie M. Petit
Robert E. Laird
Edwin L. Marsden

NEW YORK AND LONDON

First published 2010
by Routledge
711 Third Avenue, New York, NY 10017

Simultaneously published in the UK
by Routledge
2 Park Square, Milton Park, Abingdon, Oxon OX14 4RN

Routledge is an imprint of the Taylor & Francis Group, an informa business

A Focus on Fractions: Bringing Research to the Classroom (Petit, Laird, and Marsden, 2010) is a derivative product of materials created through the Vermont Mathematics Partnership funded by the National Science Foundation (EHR-0227057) and United States Department of Education (S366A020002).

Any opinions, findings, and conclusions or recommendations expressed in this material are those of the author(s) and do not necessarily reflect the views of the National Science Foundation.

Typeset in Minion by
RefineCatch Limited, Bungay, Suffolk

Library of Congress Cataloging-in-Publication Data
Petit, Marjorie M.
  A focus on fractions : bringing research to the classroom / Marjorie M. Petit, Robert E. Laird,
  Edwin L. Marsden.
    p. cm.—(Studies in mathematical thinking and learning)
  Includes bibliographical references and index.
  1. Fractions—Study and teaching (Elementary)   2. Fractions—Study and teaching (Middle
  school)   3. Ratio and proportion—Study and teaching (Elementary)   4. Ratio and proportion—
  Study and teaching (Middle school)   5. Number concept in children.   I. Laird, Robert E.
  II. Marsden, Edwin L.   III. Title.
  QA137.P47 2010
  372.7—dc22   2010001786

ISBN13: 978–0–415–80150–8 (hbk)
ISBN13: 978–0–415–80151–5 (pbk)
ISBN13: 978–0–203–85551–5 (ebk)

Dedication

This book is dedicated to the teachers and students of Vermont and Alabama who participated in Vermont Mathematics Partnership's Ongoing Assessment Project studies and scale-up and to the mathematics education researchers upon whose work this book is built. Without the foundational research by mathematics education researchers and the hundreds of interactions with Vermont and Alabama educators, this book would not have been possible.

Contents

Preface

It is safe to assume that you may have picked up this book because you are an educator who is baffled about why many students (elementary, middle school, high school, or college) have profound difficulties learning and applying fraction concepts. Be assured that you are not alone. Fractions are considered by many to be among the most difficult topics in the elementary school curriculum. As a matter of fact, in a recent national report, mathematicians and mathematics educators alike reported that problems with learning fractions interfere with learning other mathematics topics and continue to plague adults in daily tasks.

> Difficulty with learning fractions is pervasive and is an obstacle to further progress in mathematics and other domains dependent on mathematics, including algebra. It has also been linked to difficulties in adulthood, such as failure to understand medication regimens.
>
> (National Mathematics Panel Report, 2008)

A student solution like the one found in Figure i.1 underscores this point. Without an understanding of mathematics education research, one is left to wonder how it is possible that a student who can accurately add fractions can have little understanding about the magnitude of $\frac{23}{24}$.

Figure i.1

The sum of $\frac{1}{12}$ and $\frac{7}{8}$ is closest to

  A. 20
  B. 8
  C. $\frac{1}{2}$
  D. 1

Explain your answer.

$$\frac{1}{12} + \frac{7}{8} = \frac{2}{24} + \frac{21}{24} = \frac{23}{24} \quad \text{is closest to } 20.$$

This student work and all subsequent questions and student work are from VMP OGAP materials funded by the US Department of Education (Award Number S366A020002) and the National Science Foundation (Award Number EHR-0227057)

A Focus on Fractions: Bringing Research to the Classroom is designed to communicate important mathematics education research about how students develop an understanding of fractions concepts, common errors that students make, and preconceptions or misconceptions that may interfere with students learning new concepts and solving problems involving fractions. Educators have found that knowledge of these aspects of student learning related to fractions has a powerful impact on their teaching and on their students' learning.

This book grew out of a successful formative assessment project (VMP OGAP[1]) that is based on mathematics education research related to the teaching and learning of fractions. The student work samples, teacher advice, and VMP OGAP data referenced throughout the book came from interactions with hundreds of teachers and thousands of students in Vermont, Alabama, and Michigan between 2004 and 2009. Because of this, the ideas within these pages have been fostered, influenced, and practiced by countless educators in real educational settings.

Teachers in these VMP OGAP studies report that the knowledge contained in this book has helped them to:

- better understand evidence in student work;
- use the evidence to inform instruction;
- strengthen first wave instruction;
- understand the purpose of activities in their mathematics program, maximizing the potential of their instruction materials; and
- better understand fraction concepts.

The comment below is typical of how most teachers react once they gain an understanding of important mathematics education research related to how students learn fraction concepts.

> Before I used these materials I had no idea what it was about fractions that my students did not understand. It would not have occurred to me, for example, that students had a difficult time understanding that proper fractions were less than one, or that some students thought a fraction was less than zero. It was surprising to find out that one of the biggest problems for elementary students as they learn fractions is that they see proper fractions as two whole numbers and not a single value. More alarming was understanding that my instruction may have been reinforcing this idea.
>
> (VMP OGAP, personal communication, 2005)

A Book Designed for Classroom Teachers, In-service, and Pre-service Training

From its earliest inception, this has been a book for classroom teachers. Teachers will recognize that the vignettes used in various chapters describe real

---

[1] Vermont Mathematics Partnership Ongoing Assessment Project (http://vermont institutes.org/index.php/vmp/ogap)

issues that teachers face as they contemplate how best to teach fraction concepts to their students. The authentic student work used throughout the book, as well as the questions at the end of each chapter, provide teachers with numerous opportunities to analyze student thinking and to consider instructional strategies for their own students. Answers to the questions can be found at www.routledge.com/9780415801515. Each chapter also provides an opportunity for teachers to link concepts from the chapter to their own instructional materials/programs. Teachers have found these instructional links to be vital in helping to understand how their math program addresses the related research findings and in helping them to plan an effective unit of instruction.

Educators providing in-service training will also find *A Focus on Fractions: Bringing Research to the Classroom* to be a valuable component in helping their teachers become more successful teachers of fractions. In fact, several of the vignettes, the instructional links, most of the important ideas in this book, and a majority of the questions at the end of each chapter were first piloted through in-service courses. In-service leaders and participants alike found that the ideas, educational research, student work, instructional links, and end of chapter questions contained in this book provided a platform for meaningful exploration of substantive aspects of teaching fractions. Groups of math teachers from the same school as well as grade level teaching teams have found that working together around these ideas is particularly powerful.

Instructors working with pre-service teachers will find the numerous samples of student work to be valuable in bringing authentic student thinking into class discussions. In addition, pre-service teachers will be introduced to important educational research related to fractions and provided with many opportunities to "see" the research in student work, discuss research with peers, and consider the important instructional decisions that are at the heart of being an effective teacher of mathematics.

Bridging the Gap between Researchers and Practitioners

We have found that people outside the education profession are often surprised that, in general, educators are not aware of mathematics education research. In turn, teachers are surprised by two things. First, they are astonished that this research exists and that they were never exposed to it. Second, they are amazed at how readily the evidence in their students' work conforms to the findings in the research. These materials have been designed to help bridge the gap between what mathematics education researchers have discovered about the learning of fraction concepts and what teachers need to know to make effective instructional decisions.

Acknowledgments

The authors extend their appreciation to the members of the Vermont Mathematics Partnership Ongoing Assessment (OGAP) Design Team for their contributions to the research and development of OGAP materials and resources.

VMP OGAP Design Team Members:

Leslie Ercole, St. Johnsbury Middle School; Linda Gilbert, Dotham Brook School; Kendra Gorton, Milton Elementary School; Steph Hockenbury, Chamberlin School; Beth Hulbert, Barre City Elementary and Middle School; Amy Johnson, Milton Elementary School; Bob Laird, University of Vermont; Ted Marsden, Norwich University; Karen Moylan, Former VMP staff; Cathy Newton, Dotham Brook School; Susan Ojala, Vermont Mathematics Initiative, Marge Petit, Marge Petit Consulting, MPC; Regina Quinn, former director of VMP; Loree Silvis, independent consultant; Corrie Sweet, Former VMP staff; Tracy Thompson, Ottauquechee School; Jean Ward, Bennington Rutland Supervisory Union; Rebecca Young, Hardwick Schools.

We also extend our thanks to the members of the OGAP National Advisory Board for their encouragement, guidance, and advice.

OGAP National Advisory Board:

- Mary Lindquist, Callaway Professor of Mathematics Education, Emeritus; Past President of the National Council of Teachers of Mathematics
- Edward Silver, University of Michigan
- Judith Zawojewski, Illinois Institute of Technology.

Also:

Christopher Cunningham—Art Work (dinosaurs, marbles, Kelyn, Jared, egg carton, apples, scale, inch ruler, Go To and Looking Back icons).

Questions and student work samples in this book were the result of the research and development of the OGAP formative assessment system created as a part of the Vermont Mathematics Partnership funded by the National Science Foundation (EHR-0227057) and United States Department of Education (S366A020002). Any opinions, findings, and conclusions or recommendations expressed in this material are those of the author(s) and do not necessarily reflect the views of the National Science Foundation.

Modeling and Developing Understanding of Fractions

> ### Big Ideas
>
> - The use of models should permeate instruction, not be just an incidental experience, but a way of thinking, solving problems, and developing fraction concepts.
> - Students should interact with a variety of models that differ in perceptual features.
> - Modeling is a means to the mathematics, not the end.

Most teachers understand that models should be a part of fraction instruction. However, they often have many questions about their use, such as:

- What is the purpose of using models? To build fraction concepts? To use as a tool for solving problems? Other?
- My math program only uses one type of model (e.g., circle model). Is that OK?
- What is the best way to use models? For example, I am uncertain when my students should use the fraction strips they made at the beginning of the unit. Any time they want? Only for certain activities?
- Why can my students shade $\frac{3}{4}$ of a figure using an area model one day, and the very next day not be able to locate $\frac{3}{4}$ on a number line, or find $\frac{3}{4}$ of a set of objects? It just doesn't make sense to me.
- I have sixth grade students who use models to compare fractions. Is that OK? How can I move them from using a model to more efficient strategies (e.g., number sense, common denominators)?
- My textbook never provides students with an opportunity to make their own models? Is that OK?

This chapter *begins* to address some of these and other questions/issues as teachers are making instructional use of models. Subsequent chapters will show how modeling helps to build specific mathematics concepts (e.g., operations with fractions, equivalence).

The vignette that follows as well as the message that "models are the means to the mathematics, not the end" set the stage for understanding the importance of using models to help students to build an understanding of fraction concepts. While addressing some of the teacher questions about using models, it also illustrates the fine balance in using models to develop understandings without developing an over-reliance on models.

A Case Study—When Models are Used like Calculators

Mr. Smith is a fourth grade teacher who has been using the same mathematics program for the past five years. The program teaches fraction concepts through the use of only one model—the circle model. As a part of the instruction guided by this program, students make circle models representing halves, thirds, fourths, fifths, sixths, sevenths, . . ., fourteenths, which are put on display and used in all aspects of the unit. Mr. Smith has always been comfortable with using just circle models for fraction instruction.

This past year Mr. Smith participated in the Ongoing Assessment Project (OGAP) Study. He noticed that the OGAP questions did not always use the circle model, but included a variety of area models, number lines, set models, and models involving manipulatives such as pattern blocks and geoboards. However, since he was familiar with using the circle model, he charged ahead.

Midway through the unit he gave the students a question that involved comparing $\frac{3}{7}$ and $\frac{7}{8}$. The students asked if they could use their circle models on display to answer the question. Mr. Smith said they could if they needed to, but was hoping that they would not feel the need to use them.

Mr. Smith was very disappointed with what happened and was beginning to question the decision to just use circle models. With the exception of three students, all the students felt that they could **not** compare the fractions without the use of the models on the wall. He was hoping that his students would be able to visualize and justify $\frac{7}{8}$ as greater than $\frac{3}{7}$ using student drawn models or justifications based on $\frac{1}{2}$ as a benchmark. He thought that this was an easy comparison. However, instead of the models helping his students to internalize (generalize) the ideas behind the concepts, he realized that his students were using the pre-made circle models as the only way to compare fractions in the same way that students sometimes inappropriately use calculators as the only way to make calculations.

It may be that Mr. Smith's reliance on one type of model limited his students' abilities to make the important conceptual leap he intended. He was not sure. Mr. Smith realized that he needed to learn more about how to use models in his instruction and why using different models could help his students to internalize and generalize the mathematical ideas.

This vignette paints a picture of a classroom in which only one type of model (fraction circles) was used, and a classroom in which students relied on the models they made at the beginning of the unit as if they were reaching for a calculator to do a simple calculation. It may be possible that the students could compare these fractions without the pre-made models, but it was becoming clear to Mr. Smith that his students' use of models was not necessarily helping them to internalize fraction concepts in the way that he intended.

According to research Mr. Smith inadvertently made two mistakes in his use of fractions circles that may have led to his students not internalizing the concepts he intended to develop.

- His students used the fraction circles in a *"rote" way, not tied to the mathematical ideas that are embodied in the fraction circles* (Clements, 1999). This led to their dependence on the circles to compare fractions.
- He used only one model, while research suggests *that learning is facilitated when students interact with multiple models that differ in perceptual features causing students to continuously rethink and ultimately generalize the concept* (Dienes, cited in Post & Reys, 1979).

As you will see in the next section, models should be used as a way to understand and generalize mathematical ideas; that is, *models are a means to the mathematics, not the ends* (Post, 1981; Clements, 1999).

Modeling as the Means to Understanding Mathematics, Not the End

Models are mental maps mathematicians use as they solve problems or explore relationships. For example, when mathematicians are thinking about a number, they may have a number line in mind. They think about where the numbers are in relation to one another on this line, and they imagine moving back and forth along the line.

(Fosnot & Dolk, 2002, p. 73)

In this case a mathematician's model is a well-established "mental map." However, as students are developing their understanding of concepts, they will physically construct models to solve problems and represent concepts. Over time, students should move from the need always to construct or use physical models to carrying the mental image of the model, while still being able to make a model as they learn new concepts or encounter a difficult problem. This interview with Jared, a third grader, makes this case in point.

Student Interview—On Using Models

Interviewer: I know that mathematicians use models, but sometimes kids in school are uncomfortable using them.

Jared: I think it's pretty comfortable because sometimes if you try to do it in your head its gets harder and if you use like blocks or diagrams or anything it will help a lot. Sometimes my favorite thing is like a number line or a T-table or something. That's what I do a lot.

Interviewer: It's nice to hear that you are comfortable to draw or get other materials or that kind of thing.

Jared: Yeah, because it helps you do the questions a lot better.

Interviewer: Well, you can see it, right? It's not just words on a page.

Jared: Yeah, because if you do it in your head you can't do it as good. Sometimes I first use blocks. Then I sort of sometimes imagine blocks. So now I sort of do it in my head.

Interviewer: Wow!

Jared: So I can imagine blocks and I can do it without real blocks and I can do it in my head now. Because I did it with blocks and got it in my head I can do it pretty easy now.

Interviewer: Why can you do it in your head now?

Jared: Because I used blocks a lot in first and second grades, and since I did it a lot it sort of got stuck in my head.

Interviewer: What happens when a problem gets hard?

Jared: When the problems get like harder and harder, when they are really hard I sometimes need to draw or something.

(VMP, student interview, 2007)

One suspects that Jared's confidence in solving problems and using either "pictures in his head" or physical models are the result of Jared having

Figure 1.1 Jared is using a mental image to solve the problem

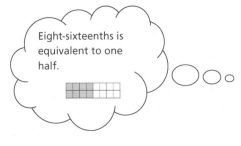

Eight-sixteenths is equivalent to one half.

experience with a variety of models over time as he developed his understanding of the concepts.

In addition, one would expect teachers to provide experiences in which students are encouraged to look for patterns and relationships, make and explore conjectures, and to use what they learn from their models to generalize concepts. Let's explore Kelyn's response to a division problem in Figure 1.2 to make this case.

Figure 1.2 Kelyn's model is a good conceptualization of this division problem

$4 \div \frac{1}{4}$ is closest to?

A. 10

B. 1

C. 0

(D. 15)

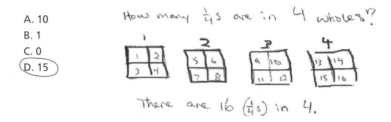

The evidence in Kelyn's solution leads one to believe that she understands that there are four one-fourths in a whole and 16 one-fourths (written in Kelyn's response as "16 $\left(\frac{1}{4}s\right)$") in four wholes. Kelyn's model is a good conceptualization of this division problem. The challenge for instruction is to ask questions or present situations that *capitalize* on this conceptualization to lead to a generalization about dividing a whole number by a proper fraction.

Here are some examples of questions that one might ask Kelyn, to help her move to a more generalized understanding of the division of fractions.

- Your model shows that there are four one-fourths in every whole. How many fourths do you think there are in 5 (or in 6 or in 10 or in 100)?
- How many thirds, fifths, or sixths are in 4 (or in 5 or in 6 or in 100)?
- What patterns do you see?
- Make and test a conjecture about the patterns that you see (giving Kelyn the chance to say "I noticed that . . .").

With additional questioning and exploration by Kelyn, she should be able to extend her understanding of unit fractions to division of whole numbers by proper fractions that are not unit fractions (Figure 1.3).

- How many three fourths are in 3 (or in 6 or in 9 or in 12)?
- How many two thirds are there in 2 (or in 4 or in 6 or in 8)?
- Make and test a conjecture about the patterns that you see.

Figure 1.3 Kelyn's mental image of $3 \div \frac{2}{3}$

Kelyn's Thinking

Well, I know that there is one two-thirds in each whole with one third left. If there are three wholes, then there are three two-thirds with three one-thirds left. Two of the thirds make another two thirds. Now I have four two-thirds and I am left with one third. One third is half of two-thirds. I think the answer is $4\frac{1}{2}$.

Jared's transition back and forth between mental models and physical models, and the potential for Kelyn's teacher to capitalize on her conceptualization of the division of fractions to make a generalization about division makes an important point. *Models are a means to the mathematics, not the end* (Post, 1981; Clements, 1999). "*The provision of multiple experiences (not the same many times) using a variety of materials, is designed to promote abstraction of the mathematical concept*" (Dienes, cited in Post, 1981).

Using models, regular probing, and asking students to explain their thinking or demonstrate their models, should play a key role in instruction as students are solving problems and building their understanding of part to whole relationships, the relative magnitude of fractions (equivalence, comparing and ordering fractions), or fraction operations.

One important point needs to be made before we proceed to the details about the features of models. *The use of models (both teacher and student generated) should permeate instruction; not just be an incidental experience, but a way of thinking and learning for students.*

- Students should have the opportunity to solve problems in which they interact with models (Figure 1.13: find $\frac{3}{8}$ of the figure);
- Students should have the opportunity to solve problems by generating their own models (Figure 1.2: Kelyn's division solution);
- Students should have the opportunity to use models to develop understanding of concepts (e.g., use a model to show that $\frac{3}{4}$ and $\frac{6}{8}$ are equivalent);

- Teachers should build on student generated models to help them generalize mathematical ideas by asking students to explain their models and respond to probing questions that capitalize on understandings in their models (e.g., Kelyn's division models).

GO TO Activities and questions related to using models to generalize concepts are found in: Chapter 4 (Partitioning)—section focusing on Using Partitioning to Generalize Concepts (p. 75); Chapter 7 (Density of Fractions)—question 2 in Looking Back; Chapter 8 (Equivalent Fractions and Comparisons)—question 2 in Looking Back; Chapter 9 (Addition and Subtraction of Fractions)—questions 2, 4, and 5 in Looking Back; Chapter 10 (Multiplication and Division of Fractions)—questions 2 and 6 in Looking Back.

Features of Area and Set Models, and Number lines

There are three different types of models that students will interact with, use to solve problems, and use to generalize concepts related to fractions—area models (regions); set models (sets of objects); and number lines. Samples of each type of model are found in Figures 1.4 to 1.6.

Using area models involves thinking about part to whole relationships. Area models that students typically interact with in mathematics programs and instruction include objects or drawings such as grids, geoboards, paper folding, and pattern blocks. Some examples of area models are pictured in Figure 1.4.

Figure 1.4 Area model samples

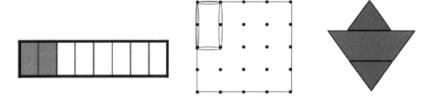

Using set models involves thinking about a fractional part of a set of objects. Set models that students typically interact with in mathematics programs and instruction include collections of common objects found in a classroom (e.g., buttons, candies, and marbles). Some examples of sets of objects are pictured in Figure 1.5.

Using a number line involves thinking about the distance traveled on a line or the location of a point on number lines, rulers, or other measurement tools. Some examples of number lines are pictured in Figure 1.6.

Figure 1.5 Set model samples—sets of objects arranged in an array (a set of apples in a 6 × 5 array), scattered (a set of 6 marbles), and in a composite set (24 eggs in 2 sets of 12 eggs)

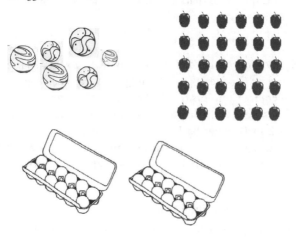

Figure 1.6 Number line from −1 to 3, a ruler, and a scale

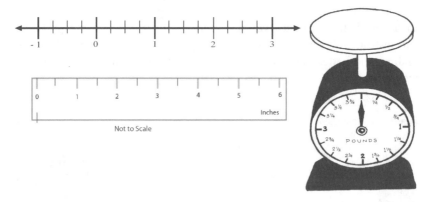

Why Models Differ in Challenges

According to research, area and set models, and number lines differ in the challenges that they present students (Hunting, cited in Bezuk & Bieck, 1993; VMP OGAP, personal communication, 2005, 2006, 2007).

Why the models differ in challenge, and why it is important for students to encounter the three types of models are related, in part, to three aspects of the models when working with fractions:

- *how the whole is defined;*
- *how "equal parts" are defined; and*
- *what the fraction indicates.*

(J. Zawojewski, personal communication, Nov. 2005)

Table 1.1 Features of area and set models, and number lines

	The whole	"Equal parts" are defined by	What the fraction indicates
Area model	The whole is determined by the area of a defined region	Equal area	The part covered of whole unit of area
Set model	The whole is determined by definition (of what is in the set)	Equal number of objects	The count of objects in the subset of the defined set of objects
Number line	Unit of distance or length (continuous)	Equal distance	The location of a point in relation to the distance from zero with regard to the defined unit

How the "Whole" and "Equal Parts" are Defined in Area and Set Models, and Number Lines with Examples

To find $\frac{1}{3}$ of the set of dinosaurs in Figure 1.8 requires different understandings than sharing a pan of brownies (Figure 1.7) equally with three people. Figures 1.7 and 1.8 illustrate how the wholes and parts in area and set models differ.

Figure 1.7 Area model—the whole is defined as an area or region (the pan of brownies). The part is defined as an equal area (although not necessarily the same shape).

The pan of brownies is the whole. When divided equally among three people, each piece is $\frac{1}{3}$ of the pan of the brownies. Each third (the part) of the pan of brownies has to be the *same area* (but doesn't require the same shape) as each of the other thirds of the pan of brownies.

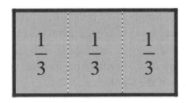

Number lines are different from area and set models. A length represents the unit which is defined by symbols (numbers), and there are continuous iterations of the unit (Figure 1.9). Number lines are used to locate a point on a line or scale designating a distance from zero, not an area of a region or a number of objects in a subset of a set.

Figure 1.8 Set model—the whole is defined as a set of objects (set of dinosaurs). The part is defined by an equal count of objects regardless of the size, color, or shape of the objects

The set of six dinosaurs is the whole. *Two* dinosaurs are $\frac{1}{3}$ of the set of six dinosaurs *regardless of the sizes* of the dinosaurs, or other differences between the dinosaurs (e.g., color, species).

Figure 1.9 A number line—the unit is defined by symbols (numbers) and the wholes (units) are continuous (connected)

A length represents the unit

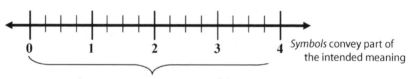

Symbols convey part of the intended meaning

There are *continuous iterations* of the unit *with simultaneous subdivisions* of the units

A key difference between an area model and a number line is that the wholes in a number line are continuous (Figure 1.9 and 1.11) while the wholes in an area model (Figure 1.10) are physically separated.

In the case of the number line, $\frac{1}{3}$ is represented by a *location* on the line or scale indicating the distance from zero, NOT $\frac{1}{3}$ of the whole line.

Go To Chapter 6: Number Lines and Fractions contains a more detailed discussion of the instructional importance of number lines and of the difficulties that students encounter when using number lines.

Regardless of the models, (a) identifying the whole (unit), (b) considering how "equal parts" are defined, and (c) what the fraction indicates, is important for students to understand in order to use models successfully.

Figure 1.10 Three pans of brownies in which the wholes are physically separate

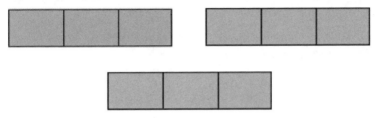

Figure 1.11 Number line from 0 to 3 in which the wholes are not physically separate, but continuous

The next two sections elaborate on other perceptual features that students encounter when interacting with models to solve problems involving fractions.

Number of Parts in the Whole

How the number of parts or objects in the whole relates to the magnitude of the denominator is another feature that needs to be considered when students solve problems involving models.

Research shows that it is easier for students to find the fractional part of the whole when the number of parts in the whole is equal to the magnitude of the denominator, than when the number of parts in the whole is a multiple or factor of the magnitude of the denominator. The most difficult case is when the number of parts in the whole is a multiple of the denominator (Bezuk & Bieck, 1993).

Figures 1.12, 1.13, 1.14, and 1.15 provide examples of the relationship of the number of parts in the whole to the denominator.

Figure 1.12 The number of parts in the whole is *equal* to the magnitude of the denominator: there are eight parts in the whole and the denominator is 8

Shade $\frac{3}{8}$ of the figure.

Teachers found that students had the most difficult time finding the fractional part of an area model when the number of parts in the whole was a multiple of the denominator and the number of rows and columns were not equal to the *denominator in Figure 1.13* (VMP OGAP, grade 4 pre- to post-student work samples, 2005).

Figure 1.13 The number of parts in the whole is a *multiple* of the denominator: there are 16 parts in the whole which is a multiple of the denominator 8

Shade $\frac{3}{8}$ of the figure.

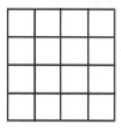

Unlike the problem in Figure 1.12, solving a problem in which the number of parts in the whole is a multiple of the denominator requires understanding the part-to-whole meaning of a fraction, and being able to divide the whole into eight equal parts. In the case of the problem in Figure 1.13, this means recognizing that $\frac{3}{8}$ of the figure means "three out of eight equal pieces." Compare Dyson's and Kim's solution to this problem in Figure 1.14.

Figure 1.14 Dyson used her part to whole understanding of $\frac{3}{8}$ and found "three out of eight *equal* parts" by first partitioning the figure into eight equal pieces and then by shading three of the equal pieces. Kim used the magnitude of the numerator and shaded three parts

Dyson's solution   Kim's solution

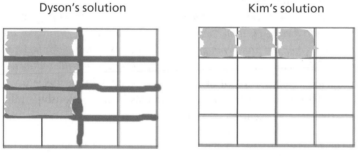

To find the fractional part of a whole when the number of parts in the whole is a factor of the denominator (Figure 1.15) requires repartitioning of the already partitioned figure. In the case of the problem in Figure 1.15, this means repartitioning the rectangle into eighths.

 For more about partitioning and part-to-whole relationships, go to Chapter 4: Partitioning, and Chapter 3: What is the Whole.

Figure 1.15 The number of parts in the whole is a *factor* of the denominator. The whole is divided into two parts which is a factor of 8

Shade $\frac{3}{8}$ of the figure.

To explore the concept of the relationship between the number of parts in the whole to the magnitude of the denominator, answer questions 1, 2, and 3 in Looking Back on p. 24.

Providing Opportunities to Interact with Models that have Different Perceptual Features

In the vignette at the beginning of this chapter, Mr. Smith used only a circle model in his instruction. This limited the students' experience with models. Other curriculum materials might use only pattern blocks or dot paper. These approaches, according to the research, would be equally limiting.

The best situation includes a balance of models *that differ in perceptual features, causing students continuously to rethink the concept (and not to overgeneralize on the strength of one model)* (Dienes, cited in Post & Reys, 1979).

When students are interacting with perceptual features of models, they are interpreting the different aspects or characteristics of the models. One way to think about this is to think about questions related to the different features of models.

Some Questions Related to Different Perceptual Features of Models

- What is the whole in the model?
- In an area model, what is the shape of the whole?
- In an area model, what is the shape of the part? How does the shape of the part relate to the shape of the whole?
- In set models, how are the objects in the set arranged (e.g., arrays, scattered)?
- In set models, are the sizes of the objects the same or different?
- Is partitioning provided or does the student have to provide the partitioning?
- What is the nature of the partitioning provided if it is provided?
- What do equal parts mean in this model?
- What is the relationship between the number of parts in the whole to the magnitude of the denominator?
- What would a student have to do to find the fractional part of the whole in question?

Look at the models in Figures 1.16 to 1.19 and think about your students interacting with each of the models for the first time. Consider the different characteristics of the models that require a "reinterpretation" as students interact and use different models.

Figure 1.16 Perceptual features of a rectangular area model

What is the whole? The area of the rectangle.
Shape of whole: Rectangle.
Shape of part: Rectangle, which may be different in size and shape.
What do equal parts mean in this model? Equal areas.

Shade $\frac{3}{4}$ of this figure.

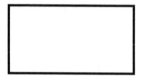

Figure 1.17 Perceptual features of an area model using pattern blocks

What is the whole? Pattern block design.
Shape of whole: Composite figure consisting of two triangles and one trapezoid.
Shape of part: Triangle, but not similar in size or shape to the whole design.
What do equal parts mean in this model? Equal areas—the size of the triangle.

What fraction of the area of John's design is made of the trapezoid?

John's design

Triangle

Trapezoid

Once teachers have understood the importance of using a variety of models with different perceptual features, they report that their instruction changes. *For example, teachers in the OGAP 2005 Exploratory Study reported:*

- *using a greater variety of models in their instruction;*
- *making explicit links between models;*
- *providing more opportunities for students to use models to solve problems;*
- *an increase in the use of number lines.*

These instructional changes appear to be reflected in student work in the OGAP 2005 Study. *Some 30 percent $\left(\frac{39}{128}\right)$ of grade 4 student pre- and*

Figure 1.18 Perceptual features of a set of marbles

What is the whole? The set of nine marbles.
How are the objects in the set arranged? Scattered.
Are the sizes of the objects in the set the same or different? Different.
What do equal parts mean? An equal count of the marbles regardless of the size of the marbles.

Circle $\frac{2}{3}$ of the marbles in the set of marbles below.

Figure 1.19 Perceptual features of a set of apples

What is the whole? The set of 30 apples.
How are the objects in the set arranged? Arranged in an array.
Are the sizes of the objects in the set the same or different? All the objects are the same size and shape.
What do equal parts mean? An equal count of apples.

Circle $\frac{1}{3}$ of the apples.

post-assessments were analyzed for the use of models to solve problems (VMP OGAP (2005). [Grade 4 pre- and post-assessment]. Unpublished raw data.).

- *In the pre-assessment only 23.1 percent* $\left(\frac{9}{39}\right)$ *of the students effectively used one or more models to solve problems.*
- *In the post-assessment 79.5 percent* $\left(\frac{31}{39}\right)$ *of the students effectively used one or more models to solve problems.*

A Case for Using Manipulatives

Up to this point the discussion has focused on pictorial models and sets of objects. However, *there is evidence that using manipulatives in instruction does facilitate learning of mathematics concepts and principles* (Driscoll et al., cited in Clements, 1999).

However, *researchers have found that using manipulatives does not guarantee success* (Baroody, cited in Clements, 1999; Fennema, cited in Clements, 1999), *nor "are manipulatives sufficient to guarantee meaningful learning"* (Clements, 1999, p. 46).

To maximize the impact of using manipulatives to build concepts, researchers suggest that:

- *teachers guard against using manipulatives in a rote manner* (as Mr. Smith did with the fraction circles); and
- *teachers make clear connections to the mathematical ideas embodied in the manipulative and do not "assume that the concepts can automatically be read off the manipulative".*

(Clements, 1999, p. 46)

Most instructional materials that focus on developing fraction concepts include opportunities for students to use commercially made manipulatives (e.g., Cuisenaire rods) and to construct their own manipulatives (e.g., fraction strips from paper folding (Figure 1.20)).

Figure 1.20 Paper folded to make fraction strips to represent halves, fourths, and eighths

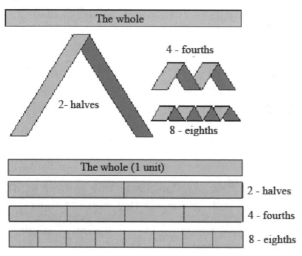

These fraction strips, for example, provide the opportunity to help students develop a number of fraction concepts *if properly used.* They can be used to help develop fraction concepts such as part-to-whole relationships as students fold the strips and compare the parts to the whole. They can be used to build equivalence understanding as students make observations comparing, say, halves, fourths and eighths—they should observe that $\frac{1}{2} = \frac{2}{4} = \frac{4}{8}$ (Figure 1.20). Fraction strips can also be used to help students make observations about the impact on partitioning that can lead to unit fraction understanding (e.g., $\frac{1}{4} > \frac{1}{8}$ because the more the paper is folded, the smaller the parts become. Therefore, one-fourth of the whole is greater than one-eighth of the same-sized whole.)

As mentioned earlier, however, for the manipulatives to be valuable the teacher plays a critical role in assuring that connections to the mathematics embodied in the manipulative are clear.

The following are some key aspects of manipulatives (student-made or commercial) that illustrate some advantages of using manipulatives:

1. Manipulatives provide an opportunity for students to move the parts in relation to the whole; or move the whole in relation to the parts— *"What is important about manipulatives is their manipulability and meaningfulness, not their physicality."*

   (Clements, cited in Viadero, April, 2007, p. 12)

2. Commercially produced manipulatives provide an opportunity for students to work with models where the size and shape of the parts and the whole are predefined. When using manipulatives *the effective use of a model to build a concept is not dependent upon a student's fine motor skills* (Lamon, 1999).

3. Different manipulatives have different perceptual features that force students to "think and rethink" their understanding of the concepts being learned.

Some commonly used manipulatives in curricular materials in addition to paper-folding include pattern blocks, geoboards, fraction bars, Cuisenaire rods, and fraction circles. Figures 1.21 and 1.22 are examples of commonly used manipulatives with descriptions of some of the perceptual features of each manipulative.

Figure 1.21 Pattern blocks

Some pattern block features when used as an area model:
- None of the pieces has the same shape or area.
- Within a given problem any piece could be used as the whole, or as the part.
- Each of the pieces is a fractional part of the hexagon, but each of the pieces is a fractional part of other parts.

Figure 1.22 Cuisenaire rods

Some Cuisenaire rod features:
- Volume model used to represent area and linear situations.
- The shape of the parts and whole are the same, but the lengths are different.
- Pieces are designed to be fractional parts of other pieces.
- Within a given problem any piece could be used as the whole or as the part.

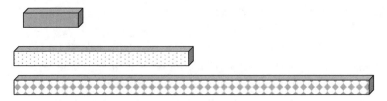

Like any strategy, the use of manipulatives as a sole strategy has its limita-tions. *Manipulatives, like Cuisenaire rods, fraction strips, and pattern blocks, while valuable because the size aspect is so closely delineated, can be limiting. "They do not allow students the freedom to break down the unit into any number of divisions"* (Lamon, 1999, p. 118).

Choosing a Model for the Situation

While it is important to use a variety of models, it is also important to recog-nize that some models can be used more effectively in some situations than in others.

Review the student responses in Figures 1.23 and 1.24 as they use models to place $\frac{1}{3}$ and $\frac{1}{4}$ in the correct locations on a number line. Both responses show evidence of effectively using a model to solve the problem.

Patty's response exemplifies a strategy that many students in the VMP 2005 OGAP Exploratory Study adopted.

It was found that student-drawn area models can be effective for making com-parisons or locating fractions on a number line when the linear feature of the model was used. However, this was only true when the lengths of the wholes were

Figure 1.23 Wesley effectively used two number lines the same size as the original number line to locate $\frac{1}{3}$ and $\frac{1}{4}$ on the number line

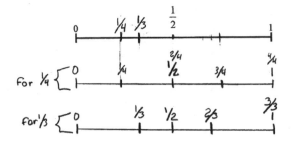

Figure 1.24 Patty effectively used the linear dimension of area models to locate $\frac{1}{3}$ and $\frac{1}{4}$ on the number line

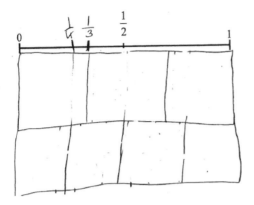

the same, and the models were partitioned into equal-sized parts (VMP OGAP, student work samples, 2005).

Using number lines and the linear feature of an area model are effective strategies for locating fractions on a number line. On the other hand, Kim's use of circles in Figure 1.25 to help locate $\frac{1}{2}$, $\frac{3}{5}$, and $\frac{2}{3}$ was not effective, given the problem situation.

Figure 1.25 Kim's response—for this situation the circle model is less effective than the area models used by Patty or the number lines used by Wesley because circle models do not translate well to the linear feature of the number line

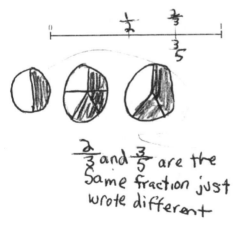

It is important to make a distinction here. Circle models can be used effectively to compare fractions as long as students consider the size of the whole and are accurate in their partitions into equal-sized parts. However, *evidence in student work has shown that circle models are not as effective in accurately locating*

fractions on a number line as when students use number lines or rectangular models (VMP OGAP, student work samples, 2005).

Student Drawn Models *May* have Limitations

While models are very important to use to develop conceptual understandings of fractions, some student-drawn models may have limitations at the younger grades when a student's fine motor skills are not developed, or when older students are comparing fractions that are close in magnitude (Lamon, 1999; VMP OGAP, student work samples, 2005). In both cases students may draw incorrect conclusions based not on lack of understanding, but on an inaccurate drawing.

Karen's response to the candy problem in Figure 1.26 provides an example about how the use of models may be limiting if the fractions are close in magnitude and the student's fine motor skills are undeveloped.

Figure 1.26 Karen's response—while the wholes in Karen's models were the same size, her inaccurate partitioning led her to the conclusion that $\frac{3}{10}$ and $\frac{2}{5}$ "are equal"

There are some candies in a dish.

$\frac{2}{5}$ of the candies are chocolate.
$\frac{3}{10}$ of the candies are peppermint.

Are there more chocolate candies or peppermint candies in the dish?

Karen incorrectly concluded that $\frac{2}{5}$ and $\frac{3}{10}$ are equal based upon her models in which the models are not partitioned into equal-sized parts. This partitioning error makes it appear that the fractions are equal.

Helping Students to Understand Modeling Errors

One way to help students understand a modeling error is to have them compare their model to a manipulative where the size of the wholes and parts are predefined.

In Figure 1.27 Leslie used area models to compare $\frac{1}{3}$ and $\frac{1}{4}$. Her model would lead to the correct conclusion that $\frac{1}{3}$ is larger than $\frac{1}{4}$. However, her modeling error (wholes not the same size) could lead to the incorrect conclusion that $\frac{2}{3}$ and $\frac{3}{4}$ are equal.

Figure 1.27 Leslie's response—because the wholes are not the same size, Leslie may conclude that $\frac{2}{3}$ and $\frac{3}{4}$ are equal

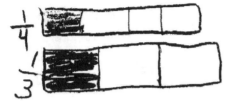

The inaccuracy of the models could be due to a misunderstanding about the importance of the wholes being the same size when comparing fractions, or it could be due to the limitations of Leslie's hand-drawn model. In either case, having Leslie compare her models to manipulatives may help her to focus on important features of models (e.g., size of the whole) to consider when comparing fractions (Figure 1.28).

Figure 1.28 Comparing manipulatives with predefined wholes and parts to student work

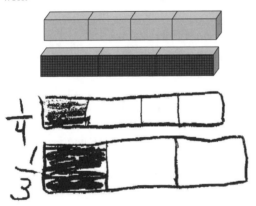

As Leslie compares the two models, the teacher should ask the student explicit questions which focus on features of the models that are necessary for the effective use of the model in making an accurate comparison of the fractions.

Possible questions using the student-drawn model and a manipulative (Figure 1.28) are:

- Which fraction is greater $\frac{2}{3}$ or $\frac{3}{4}$? Explain.
- What feature of your model leads to a different conclusion than when using the blocks? Explain.

Another way to help students improve their models is to compare different

Figure 1.29 Comparing two students' solutions—both students partitioned their models into (about) equal parts. However, the size of the wholes in Leslie's model are not the same, leading to a wrong conclusion that $\frac{3}{4}$ and $\frac{2}{3}$ are equivalent. The size of the wholes in Keisha's model are the same, leading to the correct conclusion that $\frac{3}{4} > \frac{2}{3}$

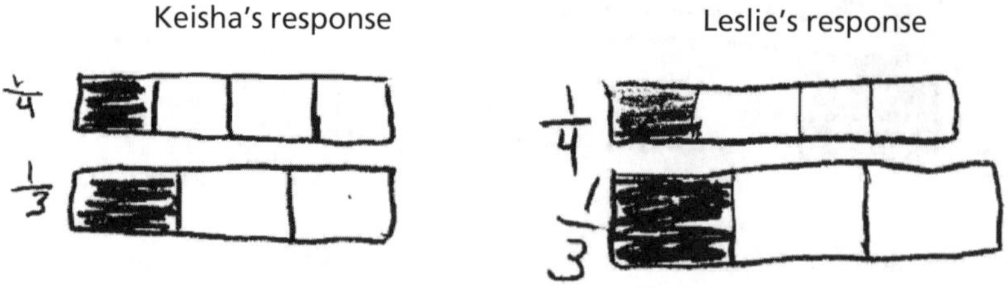

Keisha's response   Leslie's response

student solutions. Place the two solutions side by side (Figure 1.29) and then ask questions that will focus on the mathematical point being made:

- How are Keisha's and Leslie's models alike?
- How are they different?
- Using Keisha's model, it appears that $\frac{3}{4} > \frac{2}{3}$. Using Leslie's model $\frac{3}{4}$ and $\frac{2}{3}$ appear to be equal. What feature of the models led to a different conclusion?

Models are Not the Only Way to Reason with Fractions

In the vignette at the beginning of this chapter we found Mr. Smith's students having a difficult time comparing $\frac{3}{7}$ and $\frac{7}{8}$. The students were relying on the pre-made fractions circles. The students were using their fraction circles "rotely" to compare the fractions instead of using the models to internalize the mathematical ideas.

Researchers have found that students use five types of reasoning when they successfully compare and order fractions that in some way involve reasoning about the relative contributions of the numerator and denominator to the magnitude of the fractions.

- *using models;*
- *using unit fraction reasoning (fractions with numerators of one—e.g., $\frac{1}{8}$, $\frac{1}{5}$, $\frac{1}{16}$);*
- *using extended unit fraction reasoning when comparing and ordering other fractions;*
- *using a reference point like $\frac{1}{2}$; and*
- *using equivalence/common denominators.*

(Behr & Post, 1992)

GO TO These reasoning strategies are examined in more detail in Chapter 5: Comparing and Ordering Fractions and in Chapter 8: Equivalent Fractions and Comparisons.

Five Ways to Represent Mathematical Concepts—Some Implications for Fraction Instruction

The focus of this chapter has been on the use of both drawings (pictorial models) and manipulatives to represent fraction concepts. Having students interact with and develop their own models using diagrams and manipulatives are important ways to develop students' understanding of fractions.

However, *mathematics concepts can and should be represented in other ways as well: using real-world objects, spoken symbols, written words, and written symbols. According to researchers, students who experience a variety of ways to represent fractions, and are asked to move back and forth between them develop more flexible notions of fractions* (Lesh, Landau, & Hamilton, 1983).

Figure 1.30 Five-sixths represented in models, in words, in written representation, in symbolic notation, and with real-world objects help students to develop flexible notions of fractions

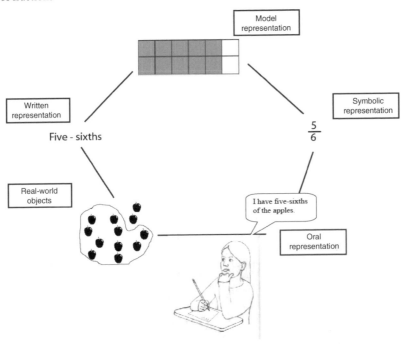

Chapter Summary

This chapter presented research related to the use of models (diagrams and manipulatives). Even though there are situations which limit the use of student-drawn models, the overall research indicates that the use of models, both pictorial and student drawn, helps students to develop a strong conceptual understanding of fractions. Furthermore, the research suggests that students should:

- interact with a variety of models in a variety of contexts to understand concepts;
- work back and forth among real-world objects, manipulatives, diagrams and pictures, spoken symbols, and written symbols; and
- use models, ultimately, to generalize mathematics concepts.

How students build understanding of different concepts using models is woven throughout the other chapters of this book.

 Looking Back

1. Compare questions in Figures 1.31 and 1.32. Data from the OGAP Exploratory study showed that students had a more difficult time with the question in Figure 1.31 than with the question in Figure 1.32. Provide a possible explanation for why one question is more difficult than the other.

Figure 1.31 Shade $\frac{1}{8}$ of this figure

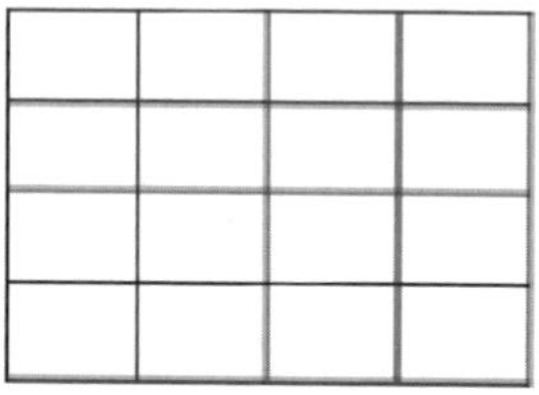

Figure 1.32 Shade $\frac{1}{8}$ of this figure

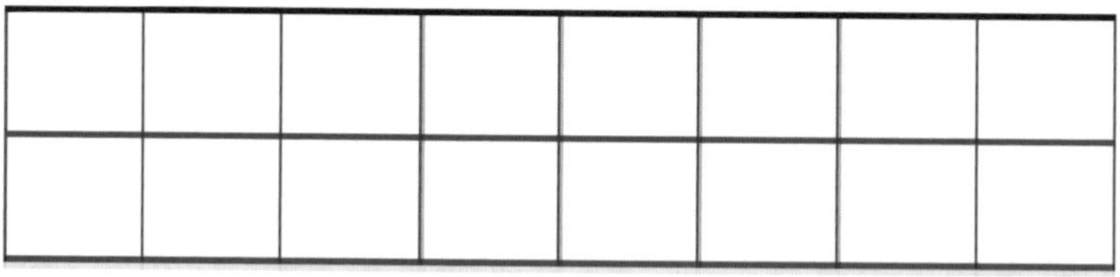

2. Why do you think that it is more difficult for a student to determine the fractional part of a whole when the number of parts in the whole is a factor or multiple of the denominator rather than when the number of parts in the whole is equal to the denominator?
3. You have just completed the first part of your fraction unit with your third-grade students. Up to this point your students have been finding the fractional part of an area as in Figure 1.33. Students have been very successful with questions like these.

Figure 1.33 Shade $\frac{3}{4}$ of this square

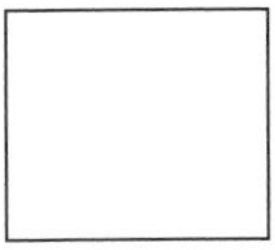

Today you are going to ask your students to find $\frac{3}{4}$ of the objects in a bag. There are four marbles and eight buttons in the bag. What aspects of the task may cause problems for your students? Explain why.

Instructional Link—Your Turn (see Table 1.2)

Table 1.2 Use the table below to help you think about how your instruction or mathematics programs provides students the opportunity to use a variety of models to solve problems, understand concepts, or generalize ideas.

What models are used in the program?	
Area	**Set**
☐ Part to whole ☐ Equivalence, comparing, ordering ☐ Operations	☐ Part to whole ☐ Equivalence, comparing, ordering ☐ Operations
Number Line	**Summary – In this program models . . .**
☐ Locating points on a number line ☐ Equivalence, comparing, ordering ☐ Operations	☐ are never used ☐ are sometimes used ☐ permeate instruction
Are students provided with the opportunity to . . .	
Answer questions in which models are given? ☐ Never ☐ Occasionally ☐ Throughout	Use manipulatives to solve problems? ☐ Never ☐ Occasionally ☐ Throughout
Use student drawn models to solve problems ☐ Never ☐ Never ☐ Never	**Use models to help develop concepts or generalize ideas?** ☐ Never ☐ Occasionally ☐ Throughout
Are there adjustments that I need to make to my instruction to assure that students experience a variety of models? If yes, describe.	

Are there adjustments that I need to make to my instruction to assure that students experience a variety of models? Describe.

Research Review—Modeling and Developing Understanding of Fractions

Researchers warn about using models in a "rote" way, not tied to the mathematical ideas that are embodied in the model (Clements, 1999).

Research suggests that learning is facilitated when students interact with multiple models that differ in perceptual features, causing students continuously to rethink and ultimately generalize the concept (Dienes, cited in Post & Reys, 1979).

> *Models are mental maps mathematicians use as they solve problems or explore relationships. For example, when mathematicians are thinking about a number, they may have a number line in mind. They think about where the numbers are in relation to one another on this line, and they imagine moving back and forth along the line.*
>
> (Fosnot & Dolk, 2002, p. 73)

Models are a means to the mathematics, not the end (Post, 1981; Clements, 1999).

"The provision of multiple experiences (not the same many times) using a variety of materials, is designed to promote abstraction of the mathematical concept" (Dienes, cited in Post, 1981).

According to research, area and set models, and number lines differ in the challenges that they present to students (Hunting, cited in Bezuk & Bieck, 1993; (VMP OGAP, personal communication, 2005, 2006, 2007).

Why the models differ in challenge, and why it is important for students to encounter the three types of models are related, in part, to three aspects of the models when working with fractions:

- *how the whole is defined;*
- *how "equal parts" are defined; and*
- *what the fraction indicates.*

(J. Zawojewski, personal communication, Nov. 2005)

Research shows that it is easier for students to find the fractional part of the whole when the number of parts in the whole is equal to the magnitude of the denominator, rather than when the number of parts in the whole is a multiple or factor of the magnitude of the denominator. The most difficult case is when the number of parts in the whole is a multiple of the denominator (Bezuk & Bieck, 1993).

Teachers found that students had the most difficult time finding the fractional part of an area model when the number of parts in the whole was a multiple of the denominator and the number of rows and columns were not equal to the denominators in Figure 1.13 (VMP OGAP, grade 4 pre- to post-student work samples, 2005).

Table 1.1 Features of area and set models, and number lines

	The whole	"Equal parts" are defined by	What the fraction indicates
Area model	The whole is determined by the area of a defined region	Equal area	The part covered of whole unit of area
Set model	The whole is determined by definition (of what is in the set)	Equal number of objects	The count of objects in the subset of the defined set of objects
Number line	Unit of distance or length (continuous)	Equal distance	The location of a point in relation to the distance from zero with regard to the defined unit

The best situation includes the use of a balance of models *that differ in perceptual features, causing students continuously to rethink the concept (and not to overgeneralize based upon one model)* (Dienes, cited in in Post & Reys, 1981).

Many teachers in the OGAP 2005 Exploratory Study (personal communication) *reported that their understanding of the importance of students interacting with models with different perceptual features led them to significant instructional change. Teachers in the OGAP 2005 Exploratory Study reported:*

- *using a greater variety of models in their instruction;*
- *making explicit links between models;*
- *providing more opportunities for students to use models to solve problems;*
- *an increase in the use of number lines.*

Some 30 percent $\left(\frac{39}{128}\right)$ of grade 4 student pre- and post-assessments were analyzed for the use of models to solve problems.

- *In the pre-assessment only 23.1 percent $\left(\frac{9}{39}\right)$ of the students effectively used one or more models to solve problems.*
- *In the post-assessment 79.5 percent $\left(\frac{31}{39}\right)$ of the students effectively used one or more models to solve problems.*

        (VMP OGAP (2005). [Grade 4 pre- and post-assessment].
                                Unpublished raw data)

There is evidence that using manipulatives in instruction does facilitate learning of mathematics concepts and principles (Driscoll, Greabell, Raphael, & Wahlstrom, Sowell, Suydam, cited in Clements, 1999).

Researchers have found that using manipulatives does not guarantee success (Baroody, cited in Clements, 1999; Fennema, cited in Clements, 1999), *nor "are manipulatives sufficient to guarantee meaningful learning"* (Clements, 1999, p. 46).

To maximize the impact of using manipulatives to build concepts, researchers suggest that:

- *teachers guard against using manipulatives in a rote manner* (as Mr. Smith did with the fraction circles); and
- *teachers make clear connections to the mathematical ideas embodied in the manipulative and not "assume that the concepts can automatically be read off the manipulative"* (Clements, 1999, p. 46).

"What is important about manipulatives is their manipulability and meaningfulness, not their physicality" (Clements, cited in Viadero, April, 2007, p. 12).

When using manipulatives *the effective use of a model to build a concept is not dependent upon a student's fine motor skills* (Lamon, 1999).

Manipulatives such as Cuisenaire rods, fraction strips, and pattern blocks, while valuable because the size aspect is so closely delineated, can be limiting. "They do not allow students the freedom to break down the unit into any number of divisions" (Lamon, 1999, p. 118).

It was found that student-drawn area models can be effective for making comparisons or locating fractions on a number line when the linear feature of the model was used. However, this was only true when the lengths of the wholes were the same, and the models were partitioned into equal-sized parts (VMP OGAP, student work samples, 2005).

Evidence in student work has shown that circle models are not as effective in accurately locating fractions on a number line as when students use number lines or rectangular models (VMP OGAP, student work samples, 2005).

While models are very important to use to develop conceptual understandings of fraction concepts, some student-drawn models may have limitations at the younger grades when a student's fine motor skills are not developed, or when older students are comparing fractions that are close in magnitude (Lamon, 1999; VMP OGAP, student work samples, 2005).

Researchers have found that students use five types of reasoning when they successfully compare and order fractions that in some way involve reasoning about the relative contributions of the numerator and denominator to the magnitude of the fractions:

- *using models;*
- *using unit fraction reasoning (fractions with numerators of one, e.g., $\frac{1}{8}$, $\frac{1}{5}$, $\frac{1}{16}$);*
- *using extended unit fraction reasoning when comparing and ordering other fractions;*
- *using a reference point such as $\frac{1}{2}$; and*
- *using equivalence/common denominators.*

(Behr & Post, 1992)

Mathematics concepts can and should be represented other ways as well: using real-world objects, spoken symbols, written words, and written symbols. According to researchers, students who experience a variety of ways to represent fractions, and are asked to move back and forth between them develop more flexible notions of fraction (Lesh, Landau, & Hamilton, 1983).

2
Inappropriate Use of Whole Number Reasoning

Big Ideas

- Students often inappropriately use whole number reasoning when solving problems involving fractions.
- Inappropriate use of whole number reasoning when solving problems involving fractions **is not inevitable.**

How do Students See this Number?

$$\frac{3}{4}$$

According to research, some students may see a fraction as two whole numbers (e.g., 3 and 4) inappropriately using whole number reasoning, not reasoning with a fraction as a single quantity (Behr, Wachsmuth, Post, & Lesh 1984; Saxe, Gearhart, Seltzer, 1999; (VMP OGAP (2005). [Grade 4 pre-assessment.] Unpublished raw data.).

Inappropriate use of *whole number reasoning* often results in students making errors when they:

- *locate fractions on a number line;*
- *compare fractions;*
- *identify fractional parts of wholes;*
- *estimate the magnitude of fractions; or*
- *operate with fractions.*

This chapter will describe the nature of inappropriate use of whole number reasoning as illustrated in student work, and will help you to think through ways in which you can assist your students to interpret fractions as quantities.

Many teachers in the OGAP studies were surprised at how readily their students inappropriately applied whole number reasoning to all aspects of fractions. In fact, a preliminary analysis of 39 fourth-grade OGAP pre-assessments $\left(\frac{39}{229}\right)$ illustrates this point. *About 44 percent $\left(\frac{137}{308}\right)$ of all incorrect responses* *analyzed on the OGAP pre-assessment were attributed to the use of inappropriate whole number reasoning* (VMP OGAP (2005). [Grade 4 pre-assessment]. Unpublished raw data.).

These teachers found that some students focused on just the numerators or on just denominators of the fraction when comparing fractions, or when finding the sums or differences of fractions, or when finding a fractional part of a whole. Figures 2.1 to 2.5 provide some examples of the ways in which inappropriate whole number reasoning is evidenced in student work.

The student work in the pre-assessment provided teachers with compelling evidence that their students were using inappropriate whole number reasoning as they solved fraction problems. What was not so clear was whether the inappropriate whole number reasoning was an artifact of the introductory instruction on fractions or was based on preconceptions carried over by the students from their study of whole numbers. In either case, these teachers now had critical information to use as they prepared and implemented their fraction unit.

Figure 2.1 Inappropriate whole number reasoning example—used the magnitude of the denominator to locate the fractions on the number line

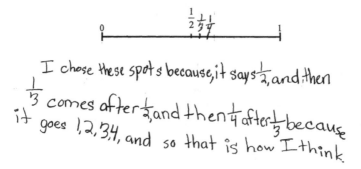

Figure 2.2 Inappropriate whole number reasoning example—used the magnitude of the numerators and the denominators to compare the fractions

There are some candies in a dish.

$\frac{2}{5}$ of the candies are chocolate.
$\frac{3}{10}$ of the candies are peppermint.

Are there more chocolate or peppermint candies in the dish?

I think there are more peppermint than chocolate because 10 is higher than 5 and 3 is also higher than 2 so I thought my answer was peppermint.

Figure 2.3 Inappropriate whole number reasoning example—circled the number of suns equal to the sum of the numerator and denominator (13 suns), not $\frac{5}{8}$ of the suns (10 suns)

Circle $\frac{5}{8}$ of the suns.

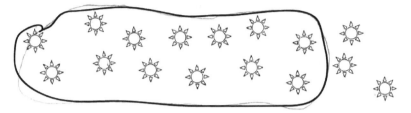

Figure 2.4 Inappropriate whole number reasoning example—added numerators and denominators to find the sum

The sum of $\frac{1}{12}$ and $\frac{7}{8}$ is closest to

    A. 20
    B. 8
    C. $\frac{1}{2}$
    D. 1

Explain your answer.

$$\frac{1}{12} + \frac{7}{8} = \frac{8}{20}$$ is close to $\frac{10}{20}$ whic his half.

Figure 2.5 Inappropriate whole number reasoning example—added sums accurately and then used the magnitude of the denominator or numerator to determine that $\frac{23}{24}$ is closest to 20

The sum of $\frac{1}{12}$ and $\frac{7}{8}$ is closest to

    A. 20
    B. 8
    C. $\frac{1}{2}$
    D. 1

Explain your answer.

$$\frac{1}{12} + \frac{7}{8} = \frac{2}{24} + \frac{21}{24} = \frac{23}{24}$$ is closest to 20.

    To help students overcome this misconception (or preconception), teachers placed a greater emphasis on comparing and ordering fractions and on the use of number lines.

    Additionally, *teachers came to recognize a situation in which they may have been inadvertently reinforcing inappropriate whole number reasoning by only providing opportunities for students to solve part-to-whole relationship problems in which the fraction in the problem has a denominator which equals the number of objects in the whole (in a set model), or the number of parts in the whole, as in the case of an area model* (VMP OGAP, personal communication, 2005). See Figure 2.6.

Figure 2.6 The number of dinosaurs in the set equals the magnitude of the denominator (six dinosaurs) and $\frac{5}{6}$ of the set of dinosaurs equals the magnitude of the numerator (five dinosaurs)

Circle $\frac{5}{6}$ of the set of dinosaurs.

This observation made by teachers points out the importance of students having experience, even in the early grades, of finding the fractional parts of a set or area where the number of objects or parts in the whole is a multiple or factor of the denominator. In particular, young students should have experience physically partitioning sets where the number of objects in the sets is a multiple of the denominator. See Figure 2.7.

Figure 2.7 Young students can physically divide sets of objects into equal groups to find a fractional part of a set of objects. Later, they can transition this understanding to paper and pencil tasks

How many dinosaurs is $\frac{1}{3}$ of a set of six dinosaurs?

Chapter Summary

Although the preponderance of inappropriate whole number reasoning was overwhelming to teachers in the OGAP pre-assessment, they realized that it was not inevitable—that their instruction mattered. Data from the OGAP 2005 Study supported their observations. *In the post-assessments 18 percent* $\left(\frac{27}{152}\right)$ *of errors were attributed to the use of inappropriate whole number reasoning compared to 44 percent* $\left(\frac{137}{308}\right)$ *in the pre-assessment* ([Grade 4 post-assessment]. Unpublished raw data.). See Table 2.1.

Table 2.1 OGAP 2005 Study—use of inappropriate whole number reasoning in OGAP pre- and post-assessments (VMP OGAP (2005). [Grade 4 pre- and post-assessment data]. Unpublished raw data.)

	Percentage of students (n = 39)	Percentage of incorrect responses	Average number of errors attributed to inappropriate whole number reasoning (only students who made error included)
Pre-assessment	85 $\left(\frac{33}{39}\right)$	44 $\left(\frac{137}{308}\right)$	4.1 (33 students)
Post-assessment	18 $\left(\frac{7}{39}\right)$	18 $\left(\frac{27}{152}\right)$	1.8 (7 students)

Teachers attributed the change from inappropriate whole number reasoning to reasoning with fractions as quantities to their increased instructional emphasis on modeling, comparing and ordering fractions, and to the use of the number line.

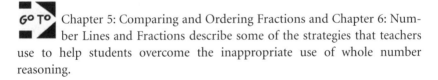 Chapter 5: Comparing and Ordering Fractions and Chapter 6: Number Lines and Fractions describe some of the strategies that teachers use to help students overcome the inappropriate use of whole number reasoning.

 Looking Back

1. Earlier in the chapter you reviewed Michael's pre-assessment response (Figure 2.1) to placing $\frac{1}{3}$ and $\frac{1}{4}$ on a number line from 0 to 1. His pre-assessment response (Figure 2.8) is shown again below with his post-assessment response to the same question (Figure 2.9). Review both responses, and then answer the following questions.

(a) What evidence in Michael's pre-assessment response suggests that Michael inappropriately used whole number reasoning when placing $\frac{1}{3}$ and $\frac{1}{4}$ on the number line?

(b) What was Michael able to do on the post-assessment that was not shown in his response on the pre-assessment?

(c) Michael's post-assessment response is very different from his pre-assessment response. What is one instructional focus that might have helped Michael to move from inappropriate use of whole number reasoning to treating each fraction as a single quantity? Use Michael's post-assessment response to support your answer?

Figure 2.8 Michael's pre-assessment response to placing $\frac{1}{3}$ and $\frac{1}{4}$ on a number line from 0 to 1

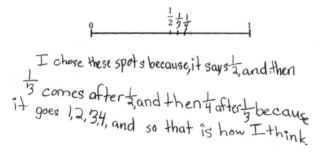

I chose these spots because, it says $\frac{1}{2}$, and then $\frac{1}{3}$ comes after $\frac{1}{2}$, and then $\frac{1}{4}$ after $\frac{1}{3}$ because it goes 1,2,34, and so that is how I think.

Figure 2.9 Michael's post-assessment response to the same question to placing $\frac{1}{3}$ and $\frac{1}{4}$ on a number line from 0 to 1

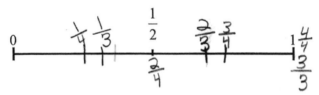

I knew that this was the correct spot because I split them in half and knew that that was $\frac{1}{4}$ and then I split them in 3rds.

2. Figures 2.10 and 2.11 include Kim's and Mark's responses to a question about the magnitude of a fraction. Both responses include diagrams generated by students. Consider their responses and then answer the following questions.

(a) What was Mark able to do? What is the evidence in Mark's response that leads one to believe that his ability to compare $\frac{3}{5}$ to a benchmark fraction is developing, but is still fragile and easily destabilized? Explain.

(b) What was Kim able to do? What is the evidence that Kim is using sound fractional reasoning? Explain.

Figure 2.10 Mark's response to $\frac{3}{5}$ is closest to

(a) 0
(b) $\frac{1}{2}$
(c) 5
(d) 8

Explain how you know.

Figure 2.11 Kim's response to $\frac{3}{5}$ is closest to

(a) 0
(b) $\frac{1}{2}$
(c) 5
(d) 8

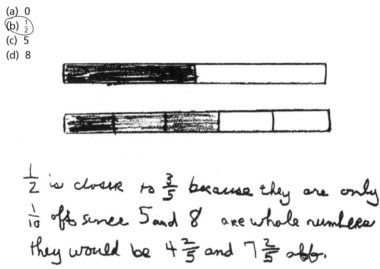

$\frac{1}{2}$ is closer to $\frac{3}{5}$ because they are only $\frac{1}{10}$ off since 5 and 8 are whole numbers they would be $4\frac{2}{5}$ and $7\frac{2}{5}$ off.

3. Review Kim's solution one more time. Kim included very carefully drawn and accurate area models for $\frac{1}{2}$ and $\frac{3}{5}$. To what extent did Kim's explanation require these area models?

4. Review the evidence in Willy's response found in Figure 2.12 and answer the questions that follow.

(a) What is the evidence in Willy's response that he had sound fractional reasoning?

(b) If Willie had the time to rewrite his response, how might his sentence be rewritten to clarify what you think Willie had in mind?

(c) Do you think that Willie decided that the sum "is just going to be a little less than 1" without computing the sum? If he didn't add the fractions, what reasoning do you think Willie used to decide that the sum "is just going to be a little less than 1?"

Figure 2.12 Willy's response

The sum of $\frac{1}{12} + \frac{7}{8}$ is closest to:

(a) 20
(b) 8
(c) $\frac{1}{2}$
(d) ①

I think 1 because $\frac{7}{8}$ is almost one + $\frac{1}{12}$ is just going to be a little less than 1.

Instructional Link—Your Turn

We suspect that some of the students in your classroom inappropriately apply whole number reasoning while solving problems involving fractions. However, this inappropriate use of whole number reasoning can be greatly reduced. If the students experience a coherent instructional program that focuses on fractions as single quantities, the students can move away from inappropriate use of whole number reasoning. To help you think about your instruction and the mathematics materials that you use, complete the following in Table 2.2.

Table 2.2 Instructional Link—strategies to support development of reasoning with fractions as quantities

Do you or your program:	Yes/no
(1) encourage students to use a variety of models in all aspects of developing fraction understanding?	
(2) provide opportunities for students to locate fractions on number lines with more than one unit? (See Chapter 6.)	
(3) provide exercises that compare fractions to benchmarks and to each other? (See Chapter 5.)	
(4) provide opportunities for students to make estimates of sums, differences, products, and quotients? (See Chapters 9 and 10.)	
(5) have a focus on reasoning with fractions as single quantities when operating with fractions, not just a focus on procedures? (See Chapters 9 and 10.)	
(6) provide the opportunity for students, even in the early grades, to find the fractional parts of a whole where the number of parts in the whole is a factor or multiple of the denominator?	

Based on the analysis above, what gaps in your instruction or mathematics program did you identify? How might you address these gaps?

Research Review—Inappropriate Use of Whole Number Reasoning

According to research, some students may see a fraction as two whole numbers (e.g., seeing 3/4 as 3 and 4) inappropriately using whole number reasoning, not reasoning with a fraction as a single quantity (Behr, Wachsmuth, Post, & Lesh 1984; Saxe, Gearhart, Seltzer, 1999; (VMP OGAP (2005). [Grade 4 pre-assessment]. Unpublished raw data.).

Inappropriate use of *whole number reasoning* often results in students making errors when they:

- *locate fractions on a number line;*
- *compare fractions;*
- *identify fractional parts of wholes;*
- *estimate the magnitude of fractions; or*
- *operate with fractions.*

About 44 percent $\left(\frac{137}{308}\right)$ of all incorrect responses analyzed on the OGAP pre-assessment were attributed to the use of inappropriate whole number reasoning (VMP OGAP (2005). [Grade 4 pre-assessment]. Unpublished raw data.).

Teachers came to recognize a situation in which they may have been inadvertently reinforcing inappropriate whole number reasoning by only providing opportunities for students to solve part-to-whole relationship problems in which the fraction of the problem has a denominator which equals the number of objects in the whole (in a set model), or the number of equal sized parts in the whole, as in the case of an area model (VMP OGAP, personal communication, 2005).

In the post-assessments 17.8 percent $\left(\frac{27}{152}\right)$ of errors were attributed to the use of inappropriate whole number reasoning compared to 44.4 percent $\left(\frac{137}{308}\right)$ in the pre-assessment (VMP OGAP (2005). [Grade 4 post assessment]. Unpublished raw data.).

Table 2.1 OGAP 2005 Study—use of inappropriate whole number reasoning in OGAP pre- and post-assessments (VMP OGAP (2005). [Grade 4 pre- and post-assessment data]. Unpublished raw data.)

	Percentage of students (n = 39)	Percentage of incorrect responses	Average number errors attributed to inappropriate whole number reasoning (only students who made error included)
Pre-assessment	85 $\left(\frac{33}{39}\right)$	44 $\left(\frac{137}{308}\right)$	4.1 (33 students)
Post-assessment	18 $\left(\frac{7}{39}\right)$	18 $\left(\frac{27}{152}\right)$	1.8 (7 students)

3
What is the Whole?

Big Ideas

- A fraction should always be interpreted in relation to the specified or understood whole.

Defining "The Whole"

"The concept of the whole underlies the concept of a fraction" (Behr & Post, 1992, p. 13). In other words, a fraction should always be interpreted in relation to the specified or understood whole.

For example, a fraction, such as $\frac{3}{4}$ has meaning only when it relates to a whole: $\frac{3}{4}$ of a set of marbles (Figure 3.1); or, $\frac{3}{4}$ of a brownie (Figure 3.2). In these cases the whole is specified.

Figure 3.1 Specified whole—the whole is the set of eight marbles. Three-fourths of the set of eight marbles is six marbles

Figure 3.2 Specified whole—the whole is the brownie, and the area model shows $\frac{3}{4}$ of the brownie remaining after one part, $\frac{1}{4}$ of the brownie, has been removed

Figure 3.3 Specified whole—the whole is the region divided into nine parts that vary in shape and size. Each part represents a fraction of the area of the whole region

Reprinted with permission from *Balanced Assessment Professional Workshop Series.* Mathematics Assessment Resource Service (MARS).

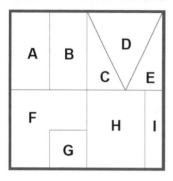

Figure 3.4 Whole not specified—the number of students (the whole) in Mrs. Smith's class and in Mr. Taylor's class is not specified, but the context implies that they are not the same

$\frac{3}{5}$ of Mrs. Smith's students ride the bus to school.
$\frac{1}{2}$ of Mr. Taylor's students ride the bus to school.
Explain how it could be possible that Mr. Taylor has more students ride the bus to school even though Mrs. Smith has a greater fractional part of her students ride the bus to school?

In other cases the whole is not specified, as in the problem in Figure 3.4.

This chapter focuses on challenges that students encounter as they develop understanding of part-to-whole relationships. Finding a fractional part of a whole is often one of the first steps in getting students to understand that fractions only have meaning in terms of a whole. Students will use these understandings to solve part-to-whole problems with larger wholes or more complex fractions (e.g., find $\frac{7}{12}$ of 144) and to develop other fraction concepts such as relative magnitude, equivalence, and operations.

Because understanding part-to-whole relationships is important, teachers should be aware of the challenges that students encounter as they solve problems involving interpreting a fraction in terms of its specified or understood whole.

- *Some students have difficulty identifying the whole when there is more than one part or object in the whole* (Payne, 1976).
- *Some students use an "out of parts strategy"* (Figure 3.8—Dominic), *not an out of "equal parts" strategy* (Figure 3.8—Abdi) *when finding the fractional part of a whole* (VMP OGAP, student work samples, 2005).
- *Some students have a difficult time determining the whole when they are*

given just a part of the whole (Behr & Post, 1992), *particularly when working with fractions that are NOT unit fractions* (Figure 3.23) (VMP OGAP, student work samples, 2005).

- *Some students make comparisons using models in which the wholes do not reflect the situation* (Figure 3.21) (VMP OGAP, student work samples, 2005).
- *Some students make comparisons using models in which the wholes are not equal in size* (Figure 3.22) (VMP OGAP, student work samples, 2005)

These are each discussed more fully later.

Identifying the Whole when there is More than One Part

Some students have difficulty identifying the whole when there is more than one part or object in the whole (Payne, 1976). This sometimes results in students *using an "out of parts strategy", not an out of equal parts strategy when finding the fractional part of a whole* (VMP OGAP, student work samples, 2005).

Tom's response (Figure 3.5), although showing one understanding of $\frac{1}{2}$, may provide evidence that he is unsure of the whole by interpreting each heart as a whole. Sonia, on the other hand, treats the set of hearts as the whole (Figure 3.6).

Figure 3.5 Tom's response—Tom circled one half of each heart

Circle $\frac{1}{2}$ (one half) of the set of hearts.

Figure 3.6 Sonia's response—Sonia correctly circled one half of the set of hearts

Circle $\frac{1}{2}$ (one half) of the set of hearts.

Another way that this problem is evidenced is when students treat an area partitioned into two parts as two wholes, as in Karen's response in Figure 3.7. Although the total area shaded in Karen's response is $\frac{1}{8}$ of the figure, Karen's use of unnecessary partitions may indicate confusion regarding the whole.

Figure 3.7 Karen's response—Karen partitioned each half-rectangle into eighths and shaded $\frac{1}{8}$ of each half using a "one out of eight" strategy

Shade $\frac{1}{8}$ of the figure below.

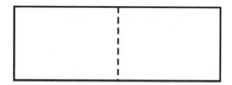

In Figure 3.8 Dominic used a "5 out 8" strategy, while Abdi is using a "5 out of 8 equal parts" strategy.

It appears that Abdi is seeing the whole as sixteen parts while Dominic is seeing two wholes of eight parts. Some teachers argue that both "answers" are correct and they are right. However, Abdi's strategy is more efficient and more "generalizable."

Consider each student solving a problem in which they have to find $\frac{5}{8}$ of $168. If Dominic used the same "out of parts" strategy, he may have to draw a figure or set of objects with 168 parts and then shade/circle "5 out of 8" 21 times. Once that is done Dominic would have to count the parts in all the shaded regions.

On the other hand, if Abdi used his "out of equal parts" strategy he might divide $168 into eight equal parts to find that each part contains $21.00. His answer then becomes $5 \times \$21.00 = \$105.00$.

Figure 3.8 Dominic shaded "5 out of 8 parts" twice. Abdi divided the 16 parts into 8 equal parts

Shade $\frac{5}{8}$ of the figure.

Dominic's response

Abdi's response

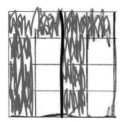

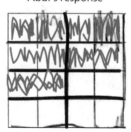

The inference from this student work and research suggests that instruction should focus on strategies to help students see the whole and to use an "out of equal parts" strategy.

One way that may help students to see the entire whole is through a familiar context in which it is not sensible to divide each object/part in the whole. This strategy is more likely to promote thinking about the whole, not each object/ part in the whole.

Examples:

- How many dinosaurs are in a $\frac{1}{2}$ of a set of 18 dinosaurs?
- Circle $\frac{1}{2}$ of a set of 20 coins.

Another strategy that helps to focus students on the whole is suggested by Lamon (1999). She suggests *that students may have an easier time identifying the whole and subsequently will make fewer partitions if they have an opportunity first to visualize the whole from a distance.* Some teachers project diagrams as shown in Figures 3.9 to 3.14. Students discuss how they visualized the fractional parts of the whole.

Figure 3.9 A set of 24 apples displayed at a distance

Figure 3.10 Some students may visualize $\frac{1}{2}$ of the set of apples, as here

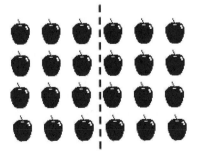

Figure 3.11 Other students may visualize $\frac{1}{2}$ of the set of apples, as here

Because the goal of projecting a set of objects is to help students visualize different ways to see the whole and potential fractional parts of the whole, select a set of objects that lends itself to exploring a range of fractional parts. For example, a set of 24 objects allows one to explore halves, thirds, fourths, sixths, eighths, twelfths, and twenty-fourths.

Figure 3.12 A student might visualize thirds

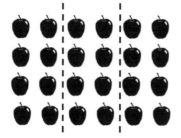

Figure 3.13 A student might visualize sixths

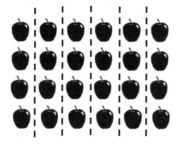

This type of visualization can also help students see equivalent fractions such as $\frac{1}{3} = \frac{2}{6}$ (Figure 3.14).

 Chapter 8: Equivalent Fractions and Comparisons for more on how to use models to develop understanding of equivalence.

Figure 3.14 One-third of the set of 24 apples is equivalent to $\frac{2}{6}$ of the set of 24 apples

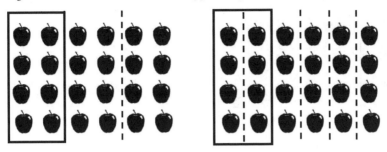

Another way to help students to focus on the whole is to provide problems such as Fractions of a Square (Figure 3.15), in which students have to re-interpret the parts, which are not the same shape or area, in terms of the whole.

Figure 3.15 Fractions of a Square

Reprinted with permission from *Balanced Assessment Professional Workshop Series*. Mathematics Assessment Resource Service (MARS).

The large outer square represents one whole unit. It has been partitioned into pieces. Each piece is identified with a letter. What fractional part of the whole is each piece?

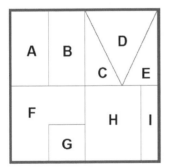

Matt's response (Figure 3.16) illustrates difficulties that students may encounter when they lose "sight" of the whole as they are solving the problem. Matt correctly wrote the fraction $\frac{1}{8}$ in the part labeled A. Matt incorrectly wrote the fraction $\frac{1}{4}$ in the part labeled I. Here part I contains $\frac{1}{4}$ of the area of the square in the lower right corner comprised of parts H and I, not $\frac{1}{4}$ of the whole square.

How can we understand Matt's selection of fractions? When Matt was look-ing at sections A and B, he used the entire square as the unit. Each of A and B contains $\frac{1}{8}$ of the area of that unit (the whole square). When Matt was looking at sections H and I, however, he incorrectly focused on only a portion of the whole figure in Figure 3.17. It appears that Matt made the same error when he considered sections F and G.

Figure 3.16 Matt's response to Fractions of a Square

Reprinted with permission from *Balanced Assessment Professional Workshop Series*. Mathematics Assessment Resource Service (MARS).

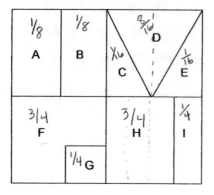

Figure 3.17 Matt only considered part (encircled) of the whole square when determining the fractional value of parts H and I

Reprinted with permission from *Balanced Assessment Professional Workshop Series*. Mathematics Assessment Resource Service (MARS).

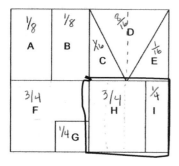

For a closer look at Matt's response to the "Fractions of a Square" problem and to consider possible next instructional steps for Matt, answer question 3 in Looking Back on p. 53

Considering the Size of the Whole when Comparing Fractions

An important concept necessary for solving problems involving fraction comparisons is to understand how the *size* of the whole impacts the fractions being considered.

Some students make comparisons using models that do not reflect the whole in the given context (VMP OGAP, student work samples, 2005).

The vignette that follows illustrates the importance of considering the whole (specified or understood) when comparing fractions.

Candy Bar

Mr. Brown is a third grade teacher. At the start of his fraction unit he always does the following activity. He comes to class with a paper bag filled with candy bars. He then says, "I have candy bars in this bag. Who would like $\frac{1}{2}$ of a candy bar and who would like a whole candy bar?"

Every year that Mr. Brown has done this activity almost all the students want a whole candy bar. The few students who don't like candy often ask for $\frac{1}{2}$ of a candy bar. Mr. Brown then hands out mini candy bars to the students who want a whole candy bar and $\frac{1}{2}$ of a large candy bar to those who only wanted $\frac{1}{2}$. Naturally, none of the students are happy!

Mr. Brown's activity makes clear that the size of the whole is critical when determining a fraction of a whole. In this case, the size of $\frac{1}{2}$ of the candy bar is dependent on the size of the whole candy bar.

The problem in Figure 3.18 is an example of a problem in which the size of the whole (i.e., the number of students in each classroom) is not specified, but the context implies that they are not the same.

Figure 3.18 Students who ride the bus to school

$\frac{3}{5}$ of Mrs. Smith's students ride the bus to school.
$\frac{1}{2}$ of Mr. Taylor's students ride the bus to school.
Explain how it could be possible that Mr. Taylor has more students ride the bus to school even though Mrs. Smith has a greater fractional part of her students ride the bus to school?

The evidence in Toni's and Samantha's responses (Figures 3.19 and 3.20) for this problem suggests that they understand that the two wholes being considered (the number of students in Mrs. Smith's class and the number of students in Mr. Taylor's class) are different sizes. To solve the problem, they each specified wholes that would prove their case.

Toni's and Samantha's solutions both reflected the situation. Contrast their solutions with Jayden's and Bill's solutions in Figures 3.21 and 3.22. Jayden solved the same problem as Samantha and Toni. However, Jayden's model of the same size and same number of parts does not reflect the situation.

Bill's response in Figure 3.22 to a problem in which the understood wholes are the same is an example of a common problem found when students use models to compare fractions. In his solution he uses different sized wholes when solving a problem in which the wholes should be the same size.

Figure 3.19 Toni's response—Toni used a set model to show that $\frac{3}{5}$ of 20 students (12 students) is less than $\frac{1}{2}$ of 26 students (13 students)

Mrs. Smiths had LESS STUDents
Say that:
Mrs. Smith has 20 Students
Mr. Taylor
Rode th Bus

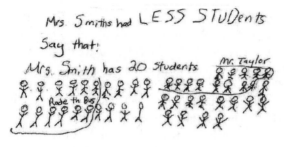

Figure 3.20 Samantha's response—Samantha did not construct models but described a situation in which Mr. Taylor had more students ride the bus to school

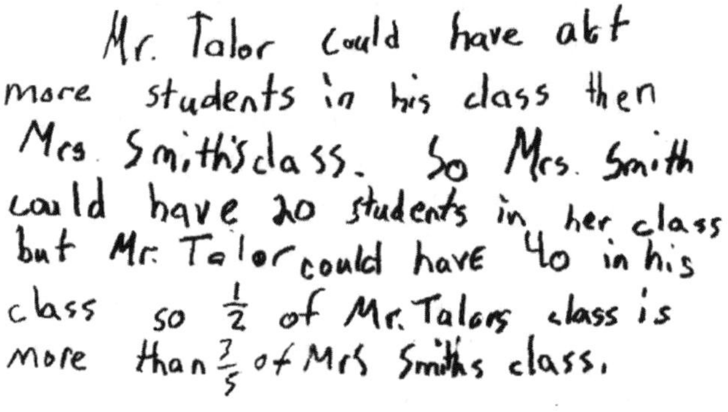

Mr. Talor could have alot
more students in his class then
Mrs. Smith's class. So Mrs. Smith
could have 20 students in her class
but Mr. Talor could have 40 in his
class so $\frac{1}{2}$ of Mr. Talors class is
more than $\frac{3}{5}$ of Mrs Smiths class.

Figure 3.21 Jayden's response—Jayden used area models to compare students who ride to school in both classrooms. However, the models and number of parts in the whole are equal while the context implies that they are not equal

$\frac{3}{5}$
$\frac{1}{2}$
Smith
taylor

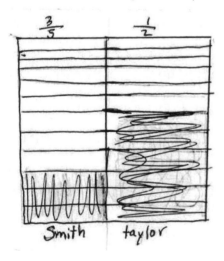

Figure 3.22 Bill's response—although Bill's models represent $\frac{5}{12}$ and $\frac{1}{3}$ they are different sizes, leading him to conclude incorrectly that $\frac{1}{3}$ of the gym is greater than $\frac{5}{12}$ of the same gym

$\frac{5}{12}$ of the gym was used for a kickball game.
$\frac{1}{3}$ of the same gym was used for a football game.

Which game used more of the gym?

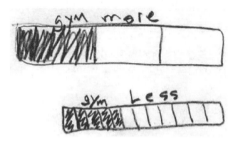

To analyze student work and consider instructional implications relative to the explicit or understood whole, complete questions 4 and 5 in Looking Back on pp. 53–55.

Given the Part, What is the Whole?

Some students also have a difficult time determining the whole when they are given just a part of the whole (Behr & Post, 1992), *particularly when dealing with fractions that are NOT unit fractions* (VMP OGAP, student work samples, 2005). Examples of student work that address this research are shown below.

Bob's responses in Figure 3.23 provide an example of a solution in which a student was successful finding the whole when given a part with a unit fraction, but not when given a non-unit fraction.

Figure 3.23 Bob's response—Bob appears to have applied the same strategy to both problems by making the number of parts in the whole equal to the magnitude of the denominator

(a) This is $\frac{1}{5}$ of a candy bar. Draw the whole candy bar.

(b) This is $\frac{7}{8}$ of another candy bar. Draw the whole candy bar.

Although Bob's diagram for part A of the problem in Figure 3.23 is "not pretty," it does show the relationship between $\frac{1}{5}$ and the whole. However, it appears that Bob applied the same strategy to both problems, treating Part B as if the problem pictured $\frac{1}{8}$ of a candy bar, not $\frac{7}{8}$ of the candy bar. It is unclear, therefore, if Bob even understands the concept for $\frac{1}{5}$. In each case he may have drawn the number of parts so that the total number of parts is equal to the magnitude of the denominator.

Contrast Bob's response to Beth's response in Figures 3.24 and 3.25. Beth's responses provide evidence that in this situation, she is able to find a whole when given a part.

Unlike Bob's response to part (b) of this problem, Beth partitioned (divided) the given part into seven equal sized pieces and then added one more piece (equal to the size of one of the seven pieces) to make a whole.

Figure 3.24 Beth's response—Beth's model shows the relationship between $\frac{1}{5}$ of the candy bar and the whole candy bar.

(a) This is $\frac{1}{5}$ of a candy bar. Draw the whole candy bar.

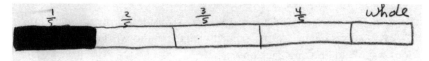

Figure 3.25 Beth's response—Beth successfully found the whole when given a nonunit fraction $\left(\frac{7}{8}\right)$

(b) This is $\frac{7}{8}$ of another candy bar. Draw the whole candy bar.

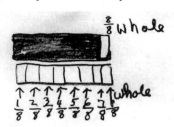

To consider next instructional steps for Bob as he deepens his understanding of identifying the whole when given a part, answer question 2 in *Looking Back* on p. 53

Chapter Summary

This chapter focused on the importance of the whole in developing fraction concepts; that is, students should solve a range of problems in which they have to think and rethink the meaning of the fraction in terms of the whole.

Students should encounter problems in which they:

- find fractional parts of wholes with multiple parts;
- compare fractions involving different sized wholes;
- compare fractions involving the same size wholes; and
- find a whole when given a part.

Looking Back

1. Explain how the lesson learned from the Candy Bar Vignette (see p. 49) was applied by the students in the *Vignette, Compare* $\frac{5}{8}$ *and* $\frac{2}{3}$, shown below.

    A group of 4th-grade students compared $\frac{5}{8}$ and $\frac{2}{3}$. As the students were presenting and discussing their solutions one student said that it didn't really matter which was bigger because $\frac{2}{3}$ is only $\frac{1}{24}$ bigger than $\frac{5}{8}$ and that wasn't very big. Another student immediately piped up and said that it depends upon the size of the whole. If the whole is really big, than $\frac{1}{24}$ could be really big (VMP OGAP, personal communication, 2005).

2. It was suggested that Bob (Figure 3.23) may have used the same strategy to solve parts (a) and (b) of the Candy Bar problem. In both cases Bob added the number of pieces that resulted in a candy bar with the total number of pieces equal to the denominator of the fraction given. Although this method resulted in a correct response to part (a), the question remains, did Bob use inappropriate whole number reasoning to solve both questions?

    What questions might you ask to determine if Bob is using inappropriate whole number reasoning and to help Bob deepen his understanding of finding the whole when given a fractional part?

3. Earlier in the chapter we examined part of Matt's work on the "Fractions of a Square" problem. Shown in Figure 3.26 is all of Matt's written work in which Matt indicated that the sum of all the parts is $\frac{20}{80}$. Matt's teacher asked him to describe this part of his solution.

    He explained, "Twenty-eightieths. $20 + 80 = 100$, that's the whole!" Use the evidence in the student work to answer the four questions (a)–(d).

    (a) What understandings of fractional parts of an area model are evidenced in Matt's response? Describe the evidence.

    (b) What errors are evidenced in Matt's response? Describe the evidence.

    (c) What potential questions might you ask Matt that would help him focus on identifying the whole?

(d) What potential questions might you ask to help Matt rethink his conclusion that $20 + 80 = 100$ and that is the whole? Provide a rationale for each question.

Figure 3.26 Fractions of a square

Reprinted with permission from *Balanced Assessment Professional Workshop Series.* Mathematics Assessment Resource Service (MARS)

The large outer square represents one whole unit. It has been partitioned into pieces. Each piece is identified with a letter. What fractional piece of the whole is each piece? Write that fraction on the piece.

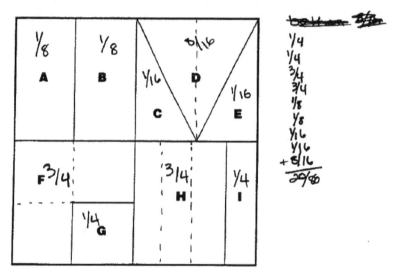

4. Read Tiara's and Maggie's responses to the "Candies in a Dish" problem in Figures 3.27 and 3.28. Although both students successfully

Figure 3.27 Tiara's response

There are some candies in a dish.

$\frac{2}{5}$ of the candies are chocolate.

$\frac{3}{10}$ of the candies are peppermint.

Are there more chocolate candies or more peppermint candies?

10 candies

$\frac{3}{10}$ of 10 candies is 3 candies and are Peppermint. $\frac{2}{5}$ of 10 candies is 4 candies that are chocolate.

there are more chocolate candies.

Figure 3.28 Maggie's response

There are some candies in a dish.

$\frac{2}{5}$ of the candies are chocolate.
$\frac{3}{10}$ of the candies are peppermint.

Are there more chocolate candies or more peppermint candies?

more chooolate candies.

answered the question, they each used a different strategy. Explain how an understanding of the whole is reflected in each of their solutions.

5. Review Jayden's work in Figure 3.29 and answer questions (a) and (b).
   (a) Jayden's solution does not reflect the situation. What is the evidence?
   (b) What are some questions that you could ask that may help Jayden rethink her solution?

Figure 3.29 Jayden's response

$\frac{3}{5}$ of Mrs. Smith's students ride the bus to school.
$\frac{1}{2}$ of Mr. Taylor's students ride the bus to school.
Explain how it could be possible that Mr. Taylor has more students ride the bus to school even though Mrs. Smith has a greater fractional part of her students ride the bus to school?

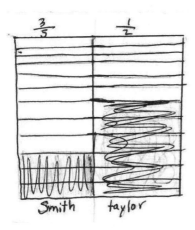

Instructional Link—Your Turn

Use the questions in Table 3.1 to help you think about how your mathematics program provides students with the opportunity to consider *the whole* and to develop understandings about the impact that the whole has on determining a fractional part of a whole.

Table 3.1 Instructional Link—strategies to support the development of solving problems involving understanding of the whole in a range of situations

Do you or does your program provide opportunities for students to:	*Yes/no*
(1) solve problems involving fractional parts of a whole?	
(2) solve problems that involve finding the fractional part of a whole that is partitioned into more than one part?	
(3) find the whole when given a part?	
(4) solve problems that involve comparing fractions that relate to the same size wholes?	
(5) solve problems that involve comparing fractions that relate to different size wholes?	
(6) construct models that reflect the size of the whole in various fraction problems and contexts?	

Describe any adjustments you need to make in your unit and lesson plans to ensure that the research from this chapter is addressed in your instruction.

Research Review—What is the whole?

"The concept of the whole underlies the concept of a fraction" (Behr & Post, 1992, p. 13).

Some students have difficulty identifying the whole when there is more than one part or object in the whole (Payne, 1976).

Some students use an "out of parts strategy" not an out of "equal parts" strategy when finding the fractional part of a whole (VMP OGAP, student work samples, 2005).

Some students have a difficult time determining the whole when they are given just a part of the whole (Behr & Post, 1992), *particularly when working with fractions that are NOT unit fractions* (Figure 3.23) (VMP OGAP, student work samples, 2005).

Some students make comparisons using models reflecting the whole that do not reflect the situation (Figure 3.21) (VMP OGAP, student work samples, 2005).

Some students make comparisons using models reflecting the whole that are not equal in size (Figure 3.22) (VMP OGAP, student work samples, 2005).

Some students have difficulty identifying the whole when there is more than one part or object in the whole (Payne, 1976). This sometimes results in students using an "out of parts strategy," not an out of equal parts strategy when finding the fractional part of a whole* (VMP OGAP, student work samples, 2005).

Another strategy that helps to focus students on the whole is suggested by Lamon (1999). She suggests *that students may have an easier time identifying the whole and subsequently will make fewer partitions if they have an opportunity first to visualize the whole from a distance.*

Some students make comparisons using models that do not reflect the whole in the given context (VMP OGAP, student work samples, 2005).

Some students also have a difficult time determining the whole when they are given just a part of the whole (Behr & Post, 1992), *particularly when dealing with fractions that are NOT unit fractions* (VMP OGAP, student work samples, 2005).

4
Partitioning

Big Ideas

Partitioning is key to understanding and generalizing concepts related to fractions:

- part-to-whole relationships;
- ordering and comparing fractions;
- equivalence;
- density of fractions; and
- operating with fractions.

What is Partitioning?

Partitioning is the act of dividing. Key to using models (area models, sets of objects, and number lines) to develop and generalize fraction concepts is the ability to partition or divide models into equal-sized parts. A partition of any of these models will separate the models into sections that "don't overlap" with each other. When each section represents the same fractional part of the whole, the sections will all be the "same size." To understand these concepts, we will explore some examples: an area model (rectangles) (Figure 4.1), a set of objects (Figure 4.2), and a number line (Figure 4.3).

Figure 4.1 Each of these rectangles is partitioned into sections that do not "over-lap." Within each rectangle, each section has the same area representing $\frac{1}{4}$ of the whole region

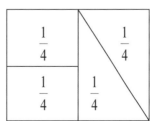

Figure 4.2 This set of eight dinosaurs has been partitioned (divided) into fourths (four groups of an equal count of dinosaurs). Each group of two dinosaurs is $\frac{1}{4}$ of the set of eight dinosaurs

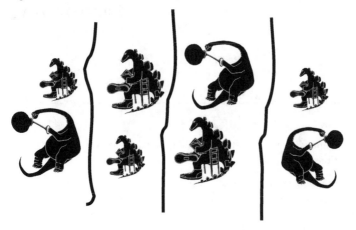

Figure 4.3 In the number line displayed below, the whole is taken to be the section of the number line from 0 to 1. This whole has been partitioned into three parts (A, B, and C), each of the same length of $\frac{1}{3}$

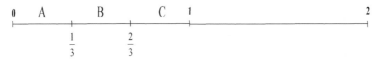

The Importance of Partitioning in Developing Fraction Concepts

Partitioning is a "fundamental mechanism for building up fraction concepts" and is key to understanding and generalizing concepts related to fractions, such as:

- *identifying "fair shares";*
- *identifying fractional parts of an object;*
- *identifying fractional parts of sets of objects;*
- *comparing and ordering fractions;*
- *locating fractions on number lines;*
- *understanding the density of rational numbers;*
- *evaluating whether two fractions are equivalent or finding equivalent fractions;*
- *operating with fractions; and*
- *measuring.* (Lamon, 1999)

Some researchers indicate that "... *early experiences with physically partitioning objects or sets of objects may be as important to a child's development of fraction concepts as counting is to their development of whole number concepts*" (Behr & Post, 1992, p. 14).

The goal, however, according to the research, *is for students to use early experiences with physically partitioning wholes to understand the impact of partitioning as they solve problems and generalize fraction concepts. Later, "just imagining the impact of partitioning will suffice and ultimately be desirable"* (Behr & Post, 1992).

Students can develop an understanding of the properties of fractions by creating partitions within models, either models that they draw or manipulatives. At a higher level of understanding, students are able to recognize how partitioning applies in solving a problem even if the model and its partition are not explicit in the students' responses. Before we begin illustrating ways that students are explicit in their use of partitioning to solve problems and ways that students solve these problems through "partitioning in their heads," read the following vignette, which highlights some important points.

Researchers indicate that *some students, when using or interpreting an area model to represent a fraction, do not consider the sizes (areas) of the different parts that result from the partitioning. Instead, the students consider just the number of parts* (Bezuk & Bieck, 1993).

The vignette also highlights another misunderstanding that students have when they partition regions: that *partitions must result in pieces that are both the same size and the same shape. Students have difficulty recognizing fractional parts as equal in size if the pieces are not congruent (same size and shape)* (Bezuk & Bieck, 1993).

Is $\frac{1}{4}$ of the square shaded?

On the day that Mrs. Armstrong was going to start a unit on fractions with her third-grade students, she introduced the lesson with the problem in Figure 4.4. She had learned that in the 2004 OGAP study only 30

Figure 4.4 Problem Mrs. Armstrong asked—is $\frac{1}{4}$ of the square shaded?

percent of third-grade students ($n = 127$) correctly responded to this problem (VMP OGAP (2004). [Grade 3 pre-assessment]. Unpublished raw data.). She knew these students well as they had been in her second-grade class during the previous year. With that second-grade class she had paid careful attention to her lessons on fractions, giving her students many experiences in which they partitioned regions into equal-sized parts. She was confident that her students would be more successful than the students in the OGAP Study.

Only 7 of her 19 students answered the question correctly. She was shocked. Many of her students shared a misinterpretation of the fractional part of an area model even after she had devoted significant instructional time to reinforcing the concept of equal-sized parts when the students were in second grade.

Correct responses to her problem looked much like Maria's, shown in Figure 4.5. These responses pleased Mrs. Armstrong and were what she had expected from all her students. It was a surprise to her that the majority of her students' solutions were incorrect. William's response, shown in Figure 4.6, was typical of these. He and many others had considered only the number of parts and not the size of the parts.

Mrs. Armstrong decided to probe a little deeper so that she could understand what her students were thinking. She asked William about his answer.

Figure 4.5 Maria's response—Maria appropriately considered the size of the pieces

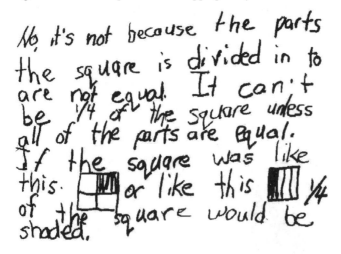

Figure 4.6 William's response to the problem in Figure 4.4—William inappropriately considered only the number of pieces

Yes because the 4 in ¼ is how many pieces there are and there are four pieces. The 1 in ¼ is how many pieces are shaded and 1 piece is shaded.

William took four marbles out of his desk to help him explain his thinking. He said that even though the marbles were not the same size, one of these marbles was $\frac{1}{4}$ of his set of four marbles—and the shaded portion of the square is also one piece out of four pieces, so it is $\frac{1}{4}$ of the square. See Figure 4.7.

Mrs. Armstrong was encouraged that her question to William had prompted his clear explanation of his thinking. His response suggested to her that some of her students had taken a feature of the fractional part in a set of objects and incorrectly used that feature in an area model.

This interaction with William raised Mrs. Armstrong's curiosity about Maria's understanding. The two area models that Maria included in her response clearly showed $\frac{1}{4}$ of the squares had been shaded. Moreover, the squares were fitted into a statement that is correct: "If the square was like (either of the partitioned and shaded squares), $\frac{1}{4}$ of the square would be shaded." That statement is very clearly written.

But now Mrs. Armstrong looked back at Maria's previous statement that "It can't be $\frac{1}{4}$ of the square unless all of the parts are equal." It appeared to Mrs. Armstrong that Maria might believe that the partitions had to result in parts of the same size and the same shape (that is, all of the parts needed to be congruent) in order to contain a part representing the fraction $\frac{1}{4}$.

Figure 4.7 William used these four marbles to explain his thinking. William thought that finding $\frac{1}{4}$ of an area is the same as finding $\frac{1}{4}$ of this set of marbles

Figure 4.8 Mrs. Armstrong asked Maria if each of the figures below showed $\frac{1}{4}$ of the figure shaded

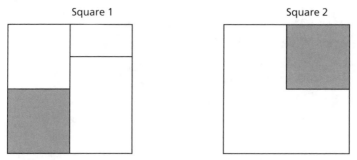

Square 1   Square 2

Mrs. Armstrong then asked Maria if each of the squares (shown in Figure 4.8) had $\frac{1}{4}$ of the square shaded.

Even though $\frac{1}{4}$ of each of the figures is shaded, Maria said, "No, $\frac{1}{4}$ is not shaded because Square 1 is not divided into four equal parts, and Square 2 is divided into two unequal parts, not four parts." Mrs. Armstrong was surprised to find that Maria did not believe that $\frac{1}{4}$ of each of the regions was shaded.

However, Maria's and William's level of understanding helped her to reflect on her instruction. While Mrs. Armstrong had emphasized the importance of partitioning regions into equal-sized parts last year, the students had never experienced questions like "Is $\frac{1}{4}$ of the square shaded?" or figures such as those in Figure 4.8.

She realized that last year the only models that students saw or drew were ones in which all the pieces were the same size and the same shape. This year she will try to be more intentional about providing her students with experiences that challenge her students' understandings of the meaning of equal-sized parts.

This vignette raises some important issues as students are introduced to and begin to use partitioning to solve problems involving finding the fractional part of a whole. Like William, students may be inappropriately applying their understanding of partitioning sets to partitioning an area model. For example, Mrs. Armstrong was realizing that William appears to understand that the objects in a set did not have to be the same size (different-sized marbles), but had to have the same number of marbles in each group. See Figure 4.9.

Mrs. Armstrong realized that some students, like Maria, may be taking a literal application of "equal-sized" parts to mean that the parts need *always* to be the same size *and* the same shape.

Figure 4.9 One-fourth of the marbles is circled even though the marbles are not the same size

As teachers reinforce the concept of equal-sized parts, they need to deepen the understanding by providing examples and counterexamples as well as developing other instructional strategies. There are many strategies for teachers to choose from that reinforce the idea that fractional parts in an area model need to be the same size, but not the same shape, and that help students move away from thinking about a fractional part of an area in the same way they think about sets of objects. One such activity asks students to come up with as many different ways as they can to show $\frac{1}{2}$ of a geoboard or $\frac{1}{2}$ of a square, as shown Figure 4.10.

Figure 4.10 Even though the shading is not the same shape, $\frac{1}{2}$ of each of these figures is shaded

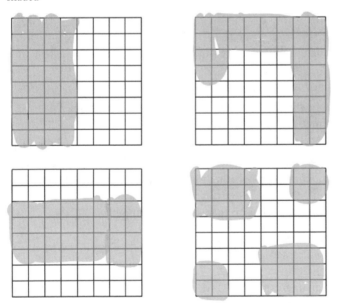

An important aspect of this activity is having students share their solutions as well as probing their understanding with questions like these.

- What fractional part of each shape is shaded? (Don't assume that all students see these as representing $\frac{1}{2}$ of each shape.)

- What did you notice about the shading on all your sketches? (For example, the shading covers the same area, but is not necessarily the same shape.)
- How did you decide what to shade? (For example, counting the total number of boxes and shading a number equal to half of the boxes.)

The examples cited in the vignette and above directly relate to solving problems involving finding the fractional part of a region (e.g., shade $\frac{1}{2}$ of a square). This concept is important as students experience real-world situations which often involve finding the fractional part of an irregular area. For example, 20 acres divided equally among four people means that all four people get five acres. It does not necessarily mean that the shape of five-acre parcels is the same. See Figure 4.11.

Figure 4.11 Each person gets $\frac{1}{4}$ of the 20 acres (or five acres). Even though the pieces are not the same shape they have the same area

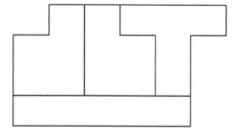

Using Partitioning and the Impact of Partitioning to Solve Problems Involving Fractions

In the last section we focused on an important concept about area models: two parts can represent the same fraction even though the parts have different shapes. What matters is that the two parts must have the same size (area).

In this section we will focus on the use of partitioning:

- to compare and order fractions;
- to operate with fractions;
- to find fair shares;
- to develop an understanding of the density of rational numbers; and
- to locate fractions on a number line.

To effectively use models to solve problems involving these mathematical topics it becomes important to partition regions into the same size and shape.

As you read through the student responses on the pages that follow, think about the following.

1. What attention are students paying to partitioning regions, lines, or sets into parts that have equal sizes?
2. How are students using their understanding of partitioning into equal-sized parts to solve the problems?
3. How are students using their understanding of the impact of partitioning to solve the problems?

Partitioning and "Fair Shares"

Students in the early grades use partitioning to divide objects into "fair shares" as in Katie's response in Figure 4.12. When first determining fair shares most students need to physically partition the region as in Katie's response. Over time and with enough experience, it is sufficient for students just to visualize the partitioning. This can be seen in Thomas's response in Figure 4.13.

Figure 4.12 Katie's response—Katie physically partitioned the pizzas into (approximately) equal-sized parts. Katie did not, however, respond with a fraction to indicate how much pizza each child got

Four children share two pizzas equally.
What fraction of a pizza does each child get?

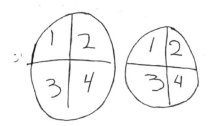

Figure 4.13 Thomas's response—Thomas used an understanding of the impact of partitioning by explaining how you can cut each pizza in half without having to draw each pizza. Using this strategy he was able to determine that each student gets $\frac{1}{2}$ of a pizza

Four children share two pizzas equally.
What fraction of a pizza does each child get?

you can cut
1 Pizza in to two halfs and
then cut the other into
two and give one half to
all of the kid's

To explore the concept of developing students' abilities to use partitioning to solve problems involving "fair shares," answer question 2, in Looking Back.

Partitioning to Compare Fractions

Students also use partitioning strategies to compare the magnitude of fractions. In Tom's response in Figure 4.14 he used a set of 12 objects to represent the class. Using two copies of this set model, he partitioned one copy into thirds and partitioned the other copy into halves. His two set models show $\frac{1}{3}$ of the set containing four children and $\frac{1}{2}$ of the set containing six children. His answer is that $\frac{1}{2}$ (of the class) is bigger and his explanation is based on counting the pieces in his partitions and saying that "6 is bigger than 4."

Figure 4.14 Tom's response—Tom understood that the number of students in the class was the same. He chose a set of 12 objects to represent the students in the class. He then physically partitioned the 12 objects into halves and thirds

$\frac{1}{3}$ of the students in Joe's class walk to school.
$\frac{1}{2}$ of the students in Joe's class ride the bus.

Do more student walk to school or ride the bus?

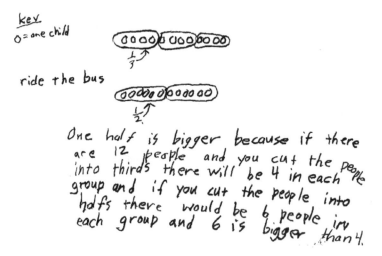

Later, students will use their understanding of the impact of partitioning on the size of the equal parts in the whole when comparing and ordering fractions as Mike did in his response in Figure 4.15.

GO TO Chapter 5: Comparing and Ordering Fractions, for a more in-depth discussion on comparing and ordering fractions.

Figure 4.15 Mike's response—Mike used his knowledge of the impact of partitioning

Linda hiked $\frac{1}{4}$ of the way up Mt. Mansfield.
Jenn hiked $\frac{1}{3}$ of the way up Mt. Mansfield. Who hiked farthest?

Jen because when the denominater is big that means the pieces are smaller. So 3 is smaller than 4 so it is $\frac{1}{3}$ since the numerator is the same.

Density of Rational Numbers

Students have a difficult time understanding the "density" of rational numbers. That is, between any two rational numbers there are an infinite number of rational numbers (Orten, Post, Behr, Cramer, Harel, & Lesh, 1995).

Experience with partitioning and repartitioning can help students to visualize the concept of density. Review Chris's response in Figure 4.16. You will

Figure 4.16 Chris's response—Chris physically partitioned two number lines in order to identify two fractions between $\frac{1}{3}$ and $\frac{3}{4}$ and showed his understanding of partitioning in (b) by indicating that "if you partition it more," you can find more fractions

(a) Name two fractions that are between $\frac{1}{3}$ and $\frac{3}{4}$.

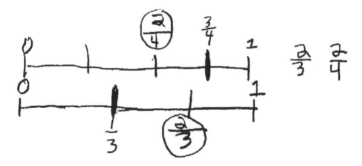

(b) Do you think that there are other fractions besides the two that you named between $\frac{1}{3}$ and $\frac{3}{4}$? Explain why or why not.

There are more in between $\frac{1}{3}$ and $\frac{3}{4}$ths because. if you partition it more you get 8ths and 6ths that and that would still — count as a fraction inbetween $\frac{1}{3}$ and $\frac{3}{4}$ths.

notice that Chris used number lines to identify two fractions $\left(\frac{2}{3} \text{ and } \frac{2}{4}\right)$ that are located between $\frac{1}{3}$ and $\frac{3}{4}$. Following his solution of part (a), Chris went on to part (b) and outlined a strategy for finding other fractions between $\frac{1}{3}$ and $\frac{3}{4}$. His response describes that you can find more fractions between $\frac{1}{3}$ and $\frac{3}{4}$ "if you partition it more."

His response here displays a comfort that Chris has with partitioning and with partitioning as a way to find fractions between two given fractions.

 GO TO Chapter 7: Density of Fractions, for a more in-depth discussion on developing student understanding of the density of fractions.

Partitioning and Operating with Fractions

Max (Figure 4.17) used partitioning strategies either to solve the problem below or to explain his solution when subtracting $\frac{2}{5}$ from $\frac{4}{5}$. Kasey (Figure 4.18),

Figure 4.17 Max's response—Max physically partitioned a number line from 0 to 1. While one does not know for sure, Max may have used the number line to subtract the two fractions or to determine that the difference $\frac{2}{5}$ was closest to 0

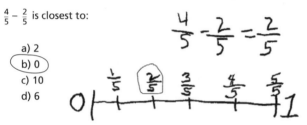

$\frac{4}{5} - \frac{2}{5}$ is closest to:

a) 2
b) 0
c) 10
d) 6

$$\frac{4}{5} - \frac{2}{5} = \frac{2}{5}$$

Figure 4.18 Kasey's response—Kasey used number sense that may have resulted from an understanding of the impact of partitioning

The sum of $\frac{1}{12} + \frac{7}{8}$ is closest to:

(a) 20
(b) 8
(c) $\frac{1}{2}$
(d) 1

$\frac{7}{8}$ is 1-8th away from 1 and 1-12th fits in between $\frac{7}{8}$ and 1.

on the other hand, was visualizing the relative magnitude of $\frac{1}{12}$ and $\frac{7}{8}$ when estimating their sum.

GO TO Chapter 9: Addition and Subtraction of Fractions, and Chapter 10: Multiplication and Division of Fractions, for a more in-depth discussion on developing student understanding of operating with fractions.

Stages of Partitioning

Since physically partioning models is such a foundational acitivity for students as they develop an understanding of fractions and generalize ideas, understanding how students develop partitioning skills is important. This next section focuses on the stages in which students develop their partitioning skills. Understanding the stages in which students develop their partitioning skills helps to provide guidance as to the possible difficulties that students might have when using partitioning to solve problems involving fractions with different denominators.

Read the following vignette.

> Mrs. Murray, a fourth-grade teacher, gave her students a pre-assessment prior to beginning a unit on fractions. She hoped to gain an insight into her students' understanding of foundational fraction concepts. She planned to use this information to inform the design and implementation of her upcoming fraction unit.
>
> She analyzed her students' responses when they were asked to place $\frac{1}{3}$ and $\frac{1}{4}$ on a 0 to 1 number line.
>
> In general, Mrs. Murray found that her students were more successful in locating and justifying the location of $\frac{1}{4}$ on a number line and less successful locating $\frac{1}{3}$. Mrs. Murray was surprised by this and wondered why her students would display a different level of understanding when placing these two fractions on the number line.

Mrs. Murray's findings are not surprising and are related to the development of partitioning strategies. *Researchers suggest that students progress through stages of partitioning which include:*

(a) sharing;
(b) algorithmic halving;
(c) evenness;
(d) oddness; and
(e) composition.

(Pothier & Sawada, 1983)

These stages are described more fully below.

Sharing

Most students first explore partitioning through sharing activities. According to research, students who successfully use a sharing strategy *are able to partition a whole into two equal parts* (Pothier & Sawada, 1983). Rhonda's work (Figure 4.19) exemplifies this sharing strategy.

Figure 4.19 Rhonda's response—Rhonda partitioned each pizza into two halves using a sharing strategy

Four children are sharing two pizzas equally.
What fraction of a pizza does each child get?

Algorithmic Halving

Students usually move easily from *sharing* to *algorithmic halving*, which is the process of *continuing the halving process to obtain fourths, eighths, sixteenths, etc.* (Pothier & Sawada, 1983). Fraction strips are used in Figure 4.20 as examples of the impact of algorithmic halving. Each fractional piece, starting with the whole strip, is *halved* to create the next smaller piece.

Partitioning models (area, set, and linear) into equal parts that are powers of two (i.e., fractions with denominators of 2, 4, 8, 16, 32, . . .) is easier than partitioning that involves odd numbers or even numbers that have odd number factors (Pothier & Sawada, 1983). This research suggests that students should be introduced to partitioning with fractions whose denominators are powers of two $\left(\frac{1}{2}, \frac{1}{4}, \frac{1}{8}, \frac{1}{16}, \ldots\right)$.

Beyond Sharing and Algorithmic Halving

Moving from partitioning that involves even numbers that are powers of two to other numbers offers students a number of challenges. Studies have shown that

Figure 4.20 Fraction strips partitioned using algorithmic halving: first, the strip is partitioned into two halves; then, each half is partitioned into halves making fourths; then, each fourth is halved, making eighths

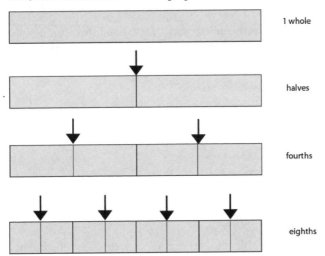

students have a more difficult time *partitioning a whole into equal parts that are odd numbers (3, 5, 7, . . .)* (Figure 4.21) *and even numbers with odd number factors* $(6 = 2 \times 3, 10 = 2 \times 5, 12 = 2 \times 2 \times 3, . . .)$ (Figure 4.22) *than partitioning using algorithmic halving strategies* (Pothier & Sawada, 1983).

Figure 4.21 Oddness stage—partitioning into an odd number *(3, 5, 7, . . .)* of equal parts involves thinking about the relative size of each part to the whole before partitioning

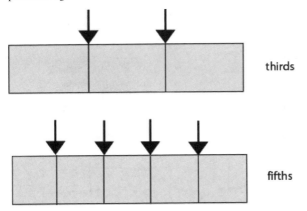

Partitioning into even numbers with odd number factors *(6, 10, 12, . . .) might involve* a two-step process: halving and partitioning into an odd number of parts. We use $6 = 2 \times 3$ to illustrate this point. One way students partition wholes into even numbers with odd number factors, such as 6, is to halve the whole first and then partition each half into thirds. Other students partition first into thirds and then halve each part. Still others will estimate and partition directly into sixths.

Figure 4.22 Evenness stage—to partition into sixths, for example, first partition the whole in half. Next, partition each half into thirds to divide the whole into sixths

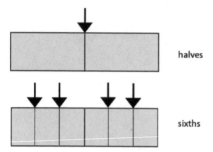

Composition

As students become flexible with the partitioning concepts described above (e.g., sharing, algorithmic halving, and partitioning into odd-number divisions), and as students' multiplicative reasoning develops, they often use multiplicative strategies to partition wholes into equal-sized parts.

Kyle and Joseph each partitioned the regions below into twelfths. Joseph's model (Figure 4.23) shows evidence of a multiplicative strategy to represent five-twelfths of the region.

Figure 4.23 Joseph's response—his strategy shows evidence of using a multiplicative strategy by partitioning the figure into 6 rows × 2 columns

 To explore the concept of stages of partitioning, answer question 2 in Looking Back on page 78.

Instruction and Stages of Partitioning

Some teachers have asked if they should explicitly teach each strategy (e.g., today we are going to do algorithmic halving). Our answer has been NO. Rather than provide steps to teach partitioning, the stages suggest that teachers make intentional choices about which fractions they use to teach, reinforce, and strengthen concepts that can be built on understanding the impact of partitioning.

To do this, a teacher should provide students with experience in which they partition a variety of models (regions, sets of objects, and number lines). Students should have experiences partitioning the models into a variety of fractional parts, starting first with even numbers that are powers of two (halves, fourths, eighths, sixteenths, . . .). To strengthen their partitioning skills a teacher should have students share their strategies so that all students are exposed to a variety of ways of thinking. Over time, students will take on other strategies as they are ready.

More importantly, students should start developing a number of generalizations from their experience with partitioning that will help them to find fractional parts of the whole, compare and order fractions, develop concepts related to equivalence, and operate with fractions.

Using Partitioning to Generalize Concepts

Teachers often ask, "How can I get my students to generalize concepts, or in this case use an understanding of the impact of partitioning?" One answer is to build from models that students use to solve problems. Another answer is to create an environment in which students are encouraged to look for patterns and relationships, ask questions, and make conjectures as Tom did in Figure 4.24.

Figure 4.24 Tom's response—Tom drew these area models to show $\frac{1}{4}$ and $\frac{1}{8}$

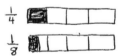

Tom made the following observation: "I noticed that the larger the denominator, the smaller the part that I shaded." He went on, "I wonder if it always works that if I make the denominator larger, I will get a smaller piece shaded?"

Tom made the following conjecture: When making a drawing to show a fractional part of a whole, the larger the denominator, the smaller the piece that will be shaded.

His teacher capitalized on his observation by engaging the class in testing Tom's conjecture. The class generated a list of the kinds of examples that can be used to test Tom's conjecture (e.g., unit fractions; the same numerator but different denominators, fractions with the same denominators, but different numerators), and then tested the conjecture with the different examples. After considerable exploration, they made more observations that led to modifying the conjecture.

Modified conjecture: When making a drawing to show a fractional part of a whole, the larger the denominator, the smaller the piece that will be shaded *if the magnitude of the numerators are the same.*

The teacher then tied this understanding into solving problems involving comparing and ordering fractions.

Read Mike's and Ted's responses below in Figures 4.25 and 4.26. These types of response exemplify responses after students move from depending on the model to solving a problem to visualizing the impact of partitioning.

Figure 4.25 Mike's response—Mike used his knoweldge of the impact of partitioning

Linda hiked $\frac{1}{4}$ of the way up Mt. Mansfield.
Jenn hiked $\frac{1}{3}$ of the way up Mt. Mansfield.
  Who hiked the farthest?

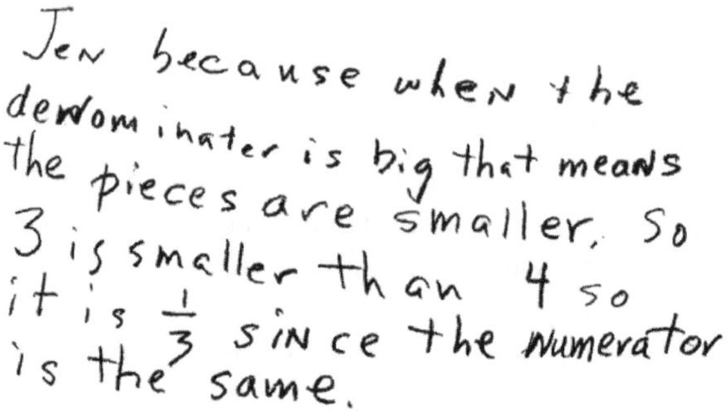

Jen because when the denominator is big that means the pieces are smaller, So 3 is smaller than 4 so it is $\frac{1}{3}$ since the numerator is the same.

Figure 4.26 Ted's response—Ted's response shows evidence of using his knowledge of the impact of partitioning to solve the problem

Which fraction is closest to 0?

$$\frac{1}{2} \quad \frac{1}{5} \quad \frac{1}{3} \quad \frac{1}{6}$$

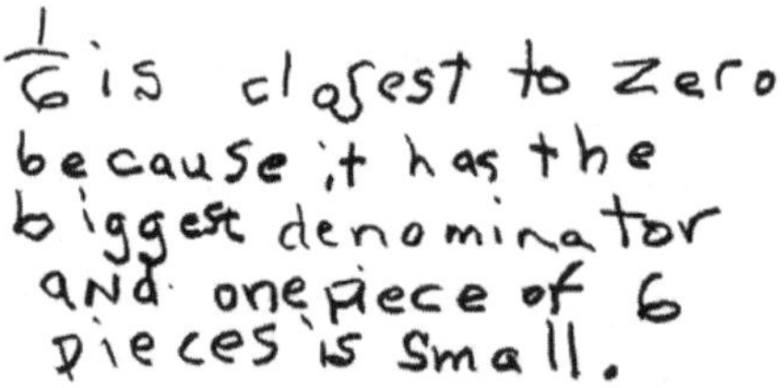

$\frac{1}{6}$ is closest to zero because it has the biggest denominator and one piece of 6 pieces is small.

We may not be sure, but it appears that the explanation that "one piece of six pieces is small" is Ted's description of what he can now visualize without drawing a model. To him the model has become a means to the mathematics, not the end.

Chapter 5: Comparing and Ordering Fractions, Chapter 7: Density of Fractions, Chapter 8: Equivalent Fractions and Comparisons, Chapter 9: Addition and Subtraction of Fractions, and Chapter 10: Multiplication and Division of Fractions provide additional opportunities to see how modeling and partitioning are used to develop and generalize concepts.

Chapter Summary

This chapter focused on how students develop skills and understanding when partitioning models, and how they use this understanding to develop and generalize fraction concepts; that is, partitioning models is a means toward understanding the mathematics. In the end we want students to solve problems and generalize concepts based on what they have internalized.

Looking Back

1. Study Mandy's and Mark's responses in Figures 4.27 and 4.28 and then answer questions (a), (b), and (c).

Figure 4.27 Mandy's response—Mandy's strategy for placing $\frac{1}{3}$ and $\frac{1}{4}$ on a number line from 0 to 1

Well half of $\frac{1}{3}$ is $\frac{1}{4}$ and Just put $\frac{1}{3}$ on after it.

Figure 4.28 Mark's strategy for locating $\frac{2}{3}$, $\frac{8}{12}$, and $\frac{8}{3}$ on the number line below

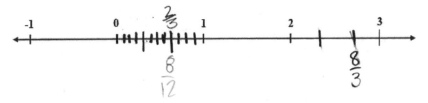

(a) What strategy does Mandy use to place $\frac{1}{4}$ on the number line? Does she use the same strategy to place $\frac{1}{3}$ on the number line? Explain, using evidence from Mandy's response.

(b) What strategy does Mark use to place $\frac{2}{3}$, $\frac{8}{12}$, and $\frac{8}{3}$ on the number line? Explain, using evidence from Mark's work.

(c) Since this is the only evidence that you have about each student's level of partitioning, what else might you want to know to determine the next instructional steps?

2. John (Figure 4.29) and Kim (Figure 4.30) answered different problems that involve dividing into "fair shares." Study their responses, and then answer questions (a) and (b).

(a) John and Kim both used partitioning in their solutions. How are their strategies different? Explain.

(b) What activity or questions might help Kim to partition each piece

Figure 4.29 John's response

Twelve students are sharing four pizzas equally.
How much will each student get?

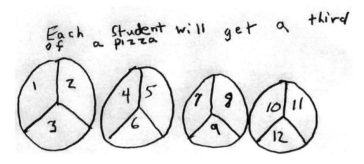

Figure 4.30 Kim's response

Six students equally share three pieces of construction paper.
How much construction paper does each child get?

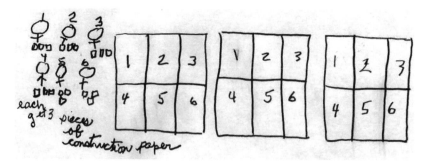

of paper into halves instead of sixths and to recognize that each student receives $\frac{1}{2}$ of a piece of construction paper?

3. Tom (Figure 4.14) and Tiara (Figure 4.31) were both asked questions about the ordering of fractions. Both of them chose to use set models in their responses. They each selected a specific number of objects in their sets, even though the numbers were not specified by the problem. Read Tom's and Tiara's responses, and answer the following questions.

   (a) Do the numbers that each selected lead to correct solutions to the problems? Explain.

   (b) Are there other numbers that Tom and Tiara could have chosen? Explain.

   (c) The solutions written by Tom and Tiara point out a unique feature of the set model as someone attempts to partition the set into parts, all of the same size. What is that unique feature? Describe.

Figure 4.31 Tiara's response

There are some candies in a dish.

$\frac{2}{5}$ of the candies are chocolate.
$\frac{3}{10}$ of the candies are peppermint.

Are there more chocolate or more peppermint candies?

$\frac{3}{10}$ of 10 candies is 3 candies that are Pepermint. $\frac{2}{5}$ of 10 candies is 4 candies that are chocolate.

there are more chocolate candies

Instructional Link—Your Turn

Use the guiding questions in Table 4.1 to help you think about how your mathematics programs provide students with the opportunity to experience partitioning and use their understanding of the impact of partitioning to generalize concepts and develop understandings of concepts.

Describe any adjustments that you need to make to your unit to ensure that students have the opportunity to use partitioning to develop an understanding of concepts as well as to generalize concepts using partitioning.

Table 4.1 Instruction Link—strategies to support the use of partitioning to develop and generalize concepts

Do you or does your program:	Yes/no
(1) provide opportunities for students to physically partition a variety of regions, sets of objects, and number lines?	
(2) pay attention to the stages of partitioning? For example, do students solve problems involving halving strategies before partitioning into an odd number?	
(3) encourage students to use their understanding of the impact of partitioning to solve problems?	
(4) use partitioning to help develop ideas or generalize concepts?	

Research Review—Partitioning

Partitioning is a "fundamental mechanism for building up fraction concepts" and is key to understanding and generalizing concepts related to fractions, such as:

- *identifying "fair shares";*
- *identifying fractional parts of an object;*
- *identifying fractional parts of sets of objects;*
- *comparing and ordering fractions;*
- *locating fractions on number lines;*
- *understanding the density of rational numbers;*
- *evaluating whether two fractions are equivalent or finding equivalent fractions;*
- *operating with fractions; and*
- *measurement.*

(Lamon, 1999)

Some researchers indicate that "... *early experiences with physically partitioning objects or sets of objects may be as important to a child's development of fraction concepts as counting is to their development of whole number concepts"* (Behr & Post, 1992).

The goal, however, according to the research, *is for students to use early experiences with physically partitioning wholes to understand the impact of partitioning as they solve problems and generalize fraction concepts. Later, "just imagining the impact of partitioning will suffice and ultimately be desirable"* (Behr & Post, 1992).

Researchers indicate that *some students, when using or interpreting an area model to represent a fraction, do not consider the sizes (areas) of the different parts that result from the partitioning. Instead, the students consider just the number of parts* (Bezuk & Bieck, 1993).

Students sometimes believe that partitions must result in pieces that are both the same size and the same shape. Students have difficulty recognizing fractional parts as equal in size if the pieces are not congruent (same size and shape) (Bezuk & Bieck, 1993).

Students have a difficult time understanding the "density" of rational numbers; that is, between any two rational numbers there are an infinite number of rational numbers (Orten et al., 1995).

Researchers suggest that students progress through stages of partitioning which include:

- *sharing;*
- *algorithmic halving;*
- *evenness;*
- *oddness; and*
- *composition.*

(Pothier & Sawada, 1983)

Partitioning is to understanding of fractions as counting is to whole numbers (Pothier & Sawada, 1983).

Students usually move easily from *sharing* to *algorithmic halving* which is the process of *continuing the halving process to obtain fourths, eighths, sixteenths, etc.* (Pothier & Sawada, 1983).

Partitioning models (area, set, and linear) into equal parts that are powers of two (i.e., fractions with denominators of 2, 4, 8, 16, 32, . . .) is easier than partitioning that involves odd numbers or even numbers that have odd number factors (Pothier & Sawada, 1983).

Partitioning a whole into equal parts that are odd numbers (3, 5, 7, . . .) and even numbers with odd number factors (6, 10, 12, . . .) is harder than partitioning using algorithmic halving strategies (Pothier & Sawada, 1983).

5
Comparing and Ordering Fractions

> ### Big Ideas
>
> - Students should develop a range of strategies for ordering and comparing fractions.

This chapter focuses on comparing and ordering positive fractions, fractions that can be written with a numerator and a denominator that are both positive. Ultimately, however, students should also be able to:

- compare a positive fraction with a negative fraction (any positive fraction is greater than any negative fraction);
- compare two negative fractions $\left(\text{e.g.,} -\frac{3}{4} < -\frac{1}{3}\right)$.

Note that many of the examples in this chapter involve making comparisons that are not in a context; e.g., which fraction is closest to 1: $\frac{1}{2}, \frac{2}{3}, \frac{3}{4}$, or $\frac{1}{5}$. Unless otherwise stated, it is assumed that the fractions being compared are associated with the same-sized whole. See Chapter 3 for more about fractions and their associated wholes.

Comparing two fractions involves determining the relative magnitude of the two fractions; that is, in a pair of fractions, are they equal to (=) each other, or is one fraction less than (<), or greater than (>) the other? (See Figure 5.1.)

Figure 5.1 Examples of comparisons of fraction pairs

$$\frac{3}{4} > \frac{1}{3}$$
$$2\frac{1}{3} < 2\frac{3}{4}$$
$$\frac{3}{4} = \frac{6}{8}$$

Ordering fractions involves putting a set of fractions in order from the least to the greatest or the greatest to the least. (See Figure 5.2.)

Figure 5.2 An example of fractions ordered from least to greatest

$$\frac{1}{6}, \frac{1}{3}, \frac{3}{8}, \frac{1}{2}, \frac{2}{3}, \frac{3}{4}$$

Study Link 7.9 (Figure 5.3) from the 2007 grade 4 Everyday Mathematics program was designed to help students to develop a range of strategies when comparing and ordering fractions.

Figure 5.3 Everyday Mathematics Study Link 7.9 Grade 4 (2007)

Wright Group/McGraw Hill, Study Link 7.9, Everyday Mathematics ®, Third Edition Grade 4 © 2007, reproduced with permission of the McGraw Hill Companies.

STUDY LINK 7·9 **Compare and Order Fractions**

Write <, >, or = to make each number sentence true.

1. $\frac{5}{6}$ _____ $\frac{1}{6}$　　　2. $\frac{3}{10}$ _____ $\frac{3}{4}$　　　3. $\frac{2}{3}$ _____ $\frac{10}{15}$

4. $\frac{10}{40}$ _____ $\frac{4}{16}$　　　5. $\frac{4}{9}$ _____ $\frac{7}{9}$　　　6. $\frac{5}{6}$ _____ $\frac{5}{8}$

7. Explain how you solved Problem 1. _____

_____

8. Explain how you solved Problem 2. _____

_____

9. Circle each fraction that is less than $\frac{1}{2}$.

$\frac{7}{8}$　　$\frac{1}{4}$　　$\frac{4}{10}$　　$\frac{7}{12}$　　$\frac{5}{9}$　　$\frac{3}{7}$　　$\frac{24}{50}$　　$\frac{67}{100}$

Write the fractions in order from smallest to largest.

10. $\frac{3}{12}$, $\frac{7}{12}$, $\frac{1}{12}$, $\frac{11}{12}$, $\frac{8}{12}$　　_____ _____ _____ _____ _____
    　　　　　　　　　　　　　　smallest　　　　　　　　　largest

11. $\frac{1}{5}$, $\frac{1}{3}$, $\frac{1}{20}$, $\frac{1}{2}$, $\frac{1}{50}$　　_____ _____ _____ _____ _____
    　　　　　　　　　　　　　　smallest　　　　　　　　　largest

12. $\frac{4}{5}$, $\frac{4}{100}$, $\frac{4}{4}$, $\frac{4}{8}$, $\frac{4}{12}$　　_____ _____ _____ _____ _____
    　　　　　　　　　　　　　　smallest　　　　　　　　　largest

Practice

13. $\frac{1}{6}$ of 30 = _____　　　14. $\frac{3}{4}$ of _____ = 75　　　15. $\frac{4}{5}$ of 45 = _____

On the surface, the page appears to be a typical "skill and drill" that involves ordering and comparing fractions that can be solved only when students know how to determine common denominators. However, that is not the case. To understand the opportunities in Study Link 7.9 we will examine:

- the relationships between the numerators and denominators of the fractions being compared or being ordered; and
- the development of a range of reasoning strategies when comparing and ordering fractions.

The Relationships between the Numerators and Denominators of the Fractions Being Compared or Ordered

Researchers indicate that students should be able to compare and order fractions which contain different combinations of numerators and denominator (classes of fraction:)

- *fractions with different numerators, but same denominators* $\left(\text{e.g.,}\right.$ for $\frac{3}{6}$ and $\frac{5}{6}, \frac{3}{6} < \frac{5}{6}\big)$;
- *fractions with the same numerators, but different denominators* $\left(\text{e.g., for}\right.$ $\frac{3}{8}$ and $\frac{3}{5}, \frac{3}{8} < \frac{3}{5}\big)$;
- *fractions with different numerators and different denominators* $\left(\text{e.g.,}\right.$ for $\frac{3}{8}$ and $\frac{5}{6}, \frac{3}{8} < \frac{5}{6}\big)$.

(Behr, Wachsmuth, Post, & Lesh, 1984)

One can find examples of each of these classes of fraction on the Study Link 7.9 (Figure 5.3). Number 1 involves comparing fractions with the same denominators. Number 2 involves solving a problem with the same numerators. Number 9 involves comparing fractions with different numerators and denominators.

Developing a Range of Reasoning Strategies when Comparing and Ordering Fractions

As you read this section on the development of reasoning by students as they compare and order fractions, think about the relationships among numerators and denominators. These relationships impact the types of reasoning that students can use as they compare and order fractions.

Researchers have found that students use five types of reasoning when they **successfully** *compare and order fractions. Each type involves determining the relative contribution of the numerator and denominator to the overall size of the fraction:*

- *reasoning with unit fractions (e.g., $\frac{1}{4} < \frac{1}{3}$ because fourths are less than thirds);*
- *extension of unit fractions reasoning to non-unit fractions (e.g., $\frac{7}{8}$ $> \frac{4}{5}$ because $\frac{1}{8}$ [the distance $\frac{7}{8}$ is away from a whole] is smaller than $\frac{1}{5}$ [the distance $\frac{4}{5}$ is away from a whole]);*
- *reasoning based on models (VMP OGAP, student work samples, 2005);*
- *reasoning through the use of a common reference fraction such as $\frac{1}{2}$; and*
- *reasoning involving equivalence.*

(Behr & Post, 1992)

Reasoning with Unit Fractions

Behr & Post (1992, p. 21) indicate that *"A child's understanding of the ordering of two fractions (that is, deciding which of the relations is equal to, is less than, or is greater than) needs to be based on an understanding of the ordering of unit fractions."*

A **unit fraction** is defined as a fraction with numerator 1 and with a denominator that is any positive whole number $\left(\text{e.g., } \frac{1}{1}, \frac{1}{2}, \frac{1}{3}, \frac{1}{4}, \ldots, \frac{1}{50}, \ldots, \frac{1}{128}, \ldots\right)$. Unit fraction reasoning involves reasoning about the relative contribution of the numerator and the denominator when only one part of the whole is considered.

Students gain an understanding of the relative magnitude of unit fractions when they interact with manipulatives and draw models to solve problems that involve comparing unit fractions.

Figure 5.4 shows the response of a student who constructed area models to compare the fractions $\frac{1}{10}$, $\frac{1}{5}$, and $\frac{1}{3}$. Teachers can capitalize on student-generated models like these by asking about the relationships in models that can lead to an understanding of the size of the parts in a whole based on the magnitude of unit fractions.

Figure 5.4 Area models partitioned into tenths, fifths, and thirds with $\frac{1}{10}$, $\frac{1}{5}$, and $\frac{1}{3}$ of each shaded. What do you notice about the size of the pieces in relationship to the magnitude of the denominator in each of the diagrams?

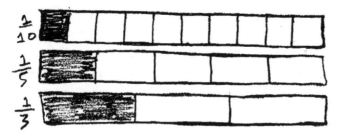

When comparing $\frac{1}{2}$, $\frac{1}{3}$, and $\frac{1}{5}$ (Figure 5.5), Kayla used unit fraction reasoning. She described the relative sizes of the denominators and then related that information to the magnitudes of the fractions by saying that "fifths are smaller parts that fourths, thirds, or halves." Her comment that "there is also only

Figure 5.5 Kayla's response—Kayla used unit fraction reasoning when comparing $\frac{1}{2}$, $\frac{1}{3}$, $\frac{1}{4}$, and $\frac{1}{5}$

Fifths are smaller parts than fourths, thirds, or halves, There is also only one part, one part" is interpreted as referring to the fact that each of the area models has one part shaded, i.e., that the fractions represented are unit fractions.

Extension of Unit Fraction Reasoning to Non-unit Fractions

Extending unit fraction reasoning means using what is known about unit fractions to compare and order fractions with:

- the same numerator, but different denominators. For example, since $\frac{1}{7}$ is less than $\frac{1}{5}, \frac{2}{7}$ must be less than $\frac{2}{5}$, or,
- different numerators and different denominators where the difference between the numerator and the denominator in each fraction is the same. For instance, the fractions $\frac{5}{7}$ and $\frac{3}{5}$ are $\frac{2}{7}$ and $\frac{2}{5}$ respectively away from the whole. Five-sevenths is two-sevenths less that one whole $\left(1-\frac{5}{7}=\frac{2}{7}\right)$ and three-fifths is two-fifths less than one whole $\left(1-\frac{3}{5}=\frac{2}{5}\right)$. Because $\frac{2}{7}<\frac{2}{5}$, subtracting $\frac{2}{7}$ is "taking away" less than subtracting $\frac{2}{5}$. The less one subtracts from 1, the closer to 1 the difference will be, thus $\frac{5}{7}>\frac{3}{5}$.

One can compare and order fractions with the same numerators but different denominators, for example $\frac{3}{8}$ and $\frac{3}{5}$, by extending their unit fraction reasoning. Using the representations of $\frac{3}{8}$ and $\frac{3}{5}$ in Figure 5.6 one can see that eighths are smaller than fifths (given the same-sized whole). Because the numerators (the number of parts shaded in each of the area models in Figure 5.6) are the same, and $\frac{1}{8}<\frac{1}{5}$, it follows that $\frac{3}{8}<\frac{3}{5}$.

Figure 5.6 The number of equal parts shaded is the same, but the size of each of the parts is not

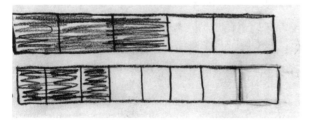

Mike's solution (Figure 5.7) contains evidence of extending his unit fraction reasoning. He describes the impact of the magnitude of the denominator on the size of the pieces. He then uses the size of one piece $\left(\frac{1}{7}\text{ and }\frac{1}{5}\right)$ to compare $\frac{3}{7}$ and $\frac{3}{5}$. Finally, he concludes that "because they $\left(\frac{3}{7}\text{ and }\frac{3}{5}\right)$ have the same numerators."

When comparing $\frac{3}{4}$ to $\frac{2}{3}$, Sam (Figure 5.8) extends his unit fraction reasoning. He compares fractions with different numerators and different denominators where the difference between the numerators and the denominators is the same. His explanation that "Both are one part from a whole" can be rewritten as $1 - \frac{3}{4} = \frac{1}{4}$ and $1 - \frac{2}{3} = \frac{1}{3}$. He continues his explanation that "thirds are bigger pieces than fourths," that is, since thirds are bigger pieces than fourths and each fraction is one part from the whole, Sam concludes that "$\frac{3}{4}$ is closer to 1 than $\frac{2}{3}$."

Figure 5.7 Mike's response—his response is an example of extending unit fraction understanding to non-unit fractions

Linda hiked $\frac{3}{7}$ of the way up Mt. Mansfield. Jen hiked $\frac{3}{5}$ of the way up Mt. Mansfield. Who hiked the farthest?

Jenn has hiked more because when the denominator is bigger the pieces are smaller. That means that $\frac{1}{7}$ is smaller than $\frac{1}{5}$ so $\frac{3}{7}$ is smaller than $\frac{3}{5}$ because they have the same numerators.

Figure 5.8 Sam's response—which shows evidence of using extended unit fraction reasoning by first recognizing that each fraction is one fractional part from a whole and then using the relative distance each fraction is from 1

$\frac{3}{4}$ is bigger than $\frac{2}{3}$. Both are 1 part from a whole - $\frac{1}{4}$ and $\frac{1}{3}$ from a whole. Thirds are bigger pieces than fourths. $\frac{3}{4}$ is closer to 1 than $\frac{2}{3}$.

The same reasoning can be used when comparing fractions in which both fractions are two or more parts away from the whole. Using the area models in Figure 5.9 comparing $\frac{6}{8}$ to $\frac{5}{7}$, one can see that eighths are smaller pieces than sevenths. Since $\frac{1}{8} < \frac{1}{7}$, then $\frac{2}{8} < \frac{2}{7}$. Therefore, $\frac{6}{8}$ is closer to 1 than $\frac{5}{7}$.

Figure 5.9 Area model comparing $\frac{6}{8}$ to $\frac{5}{7}$

Comparing to Benchmarks

Another reasoning strategy that students can use to compare and order fractions is to use reference points. Reference points such as $0, \frac{1}{2}$, and 1 can be useful benchmarks/landmarks for students to consider as they order and compare fractions.

Hannah compared $\frac{5}{6}$ to $\frac{3}{6}$ by comparing each of the fractions to the benchmark $\frac{1}{2}$ (Figure 5.10).

Figure 5.10 Hannah's response—Hannah compared $\frac{5}{6}$ and $\frac{3}{6}$ to $\frac{1}{2}$ to determine that $\frac{5}{6}$ is greater than $\frac{3}{6}$

Sam and Don each have a garden. The gardens are the same size. $\frac{5}{6}$ of Don's garden is planted with corn. $\frac{3}{6}$ of Sam's garden is planted with corn. Who has more corn in his garden?

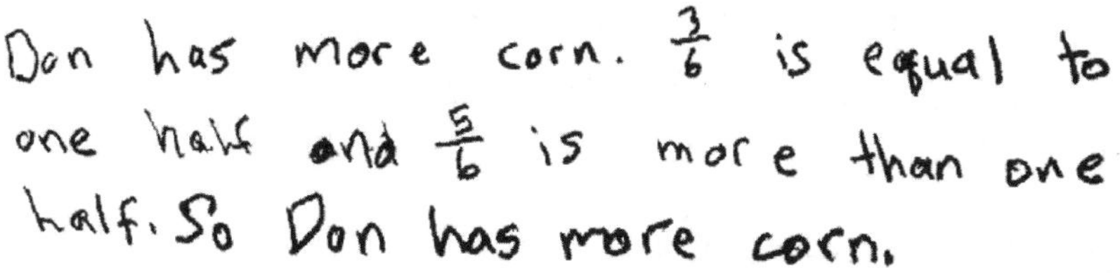

Don has more corn. $\frac{3}{6}$ is equal to one half and $\frac{5}{6}$ is more than one half. So Don has more corn.

As students become flexible in using benchmarks, they will be able to compare and order unfamiliar fractions such as $\frac{13}{24}$ and $\frac{24}{50}$, noticing that $\frac{13}{24}$ is $\frac{1}{24}$ greater than $\frac{1}{2}$ and $\frac{24}{50}$ is less than $\frac{1}{2}$ by $\frac{1}{50}$.

Equivalence—Common Denominators

Behr & Post (1992, p. 23) found that "*Ultimately the problem of ordering two general fractions (different numerators and denominators) rests on considerable knowledge of fraction equivalence.*"

GO TO Chapter 8: Equivalent Fractions and Comparisons for a discussion on how equivalence relates to the use of common denominators to compare fractions.

Inappropriately Using Whole Number Reasoning to Compare and Order Fractions

GO TO Chapter 2: Inappropriate Use of Whole Number Reasoning for a detailed discussion of inappropriate whole number reasoning and comparing and ordering fractions.

Flexibility over Time

With experience in modeling and generalizing ideas, students will move beyond their hand-drawn models and use "mental" models as they generalize ideas to flexibly compare and order fractions. For example, Nicolas used a range of reasoning strategies in Figure 5.11. Nicholas first compared each fraction $\left(\frac{1}{2}, \frac{1}{6}, \frac{11}{13}, \frac{7}{9}\right)$ to the benchmark 1. He then eliminated $\frac{1}{2}$ and $\frac{1}{6}$ as being farther from 1 than $\frac{11}{13}$ or $\frac{7}{9}$. Finally, he used his extended unit fraction reasoning to compare $\frac{11}{13}$ to $\frac{7}{9}$ —"thirteenths are smaller than ninths so $\frac{11}{13}$ is closest to 1."

Figure 5.11 Nicholas's response—Nicholas used the benchmark 1 and extended unit fraction reasoning to identify $\frac{11}{13}$ as the largest fraction in the set

Which fraction is closer to 1: $\frac{1}{2}, \frac{1}{6}, \frac{11}{13},$ or $\frac{7}{9}$?

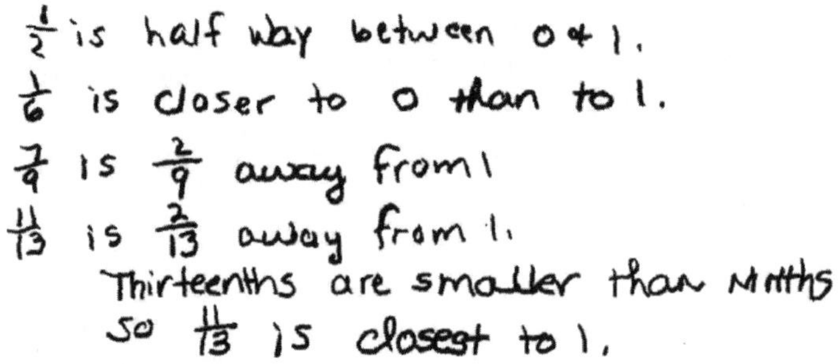

$\frac{1}{2}$ is half way between 0 & 1.

$\frac{1}{6}$ is closer to 0 than to 1.

$\frac{7}{9}$ is $\frac{2}{9}$ away from 1

$\frac{11}{13}$ is $\frac{2}{13}$ away from 1.

Thirteenths are smaller than Ninths so $\frac{11}{13}$ is closest to 1.

Teachers Need Flexibility Too

The evidence in the 2005 OGAP Exploratory Study indicated that the major strategy that students in grades 3–5 applied in the OGAP Student Post-Assessment to compare and order fractions involved the use of models. While the use of models to compare and order fractions appeared to result in a decrease in use of whole number reasoning found in the pre-assessment, developers saw few examples of students using unit fraction reasoning, extended unit fraction reasoning, equivalence ideas, or benchmarks (VMP OGAP, student work samples, 2005).

Developers suspected that the students' focus on models, not other reasoning strategies, was directly related to instruction. We decided to broaden the OGAP professional development for teachers to emphasize the use of a range of strategies as described in this chapter. To test the conjecture and measure the impact of the professional development focused on ordering and comparing fractions, teachers involved in the OGAP Scale-up (2006–7) solved the problem found in Figure 5.12.

This problem was specifically designed to provide opportunities to be solved using any of the five strategies described in this chapter with the use of a common denominator or models as the least efficient ways to solve this problem. Yet, two-thirds of the pre-assessments sampled *(n = 67)* showed evidence of using only a model or common denominators in the pre-assessment (VMP OGAP (2006). [OGAP Scale-up Pre- and Post-Assessment]. Unpublished raw data.) as exemplified in the teacher pre-assessment response in Figure 5.13.

Figure 5.12 OGAP teacher pre-assessment question

Which fraction is closest to 1: $\frac{1}{2}$, $\frac{7}{9}$, $\frac{11}{13}$, or $\frac{1}{6}$?
Solve this problem using three different strategies.

Figure 5.13 Sample OGAP teacher pre-assessment (2006–7) in which common denominators and a general description about the use of fraction bars were the only solutions used to identify the fraction closest to 1

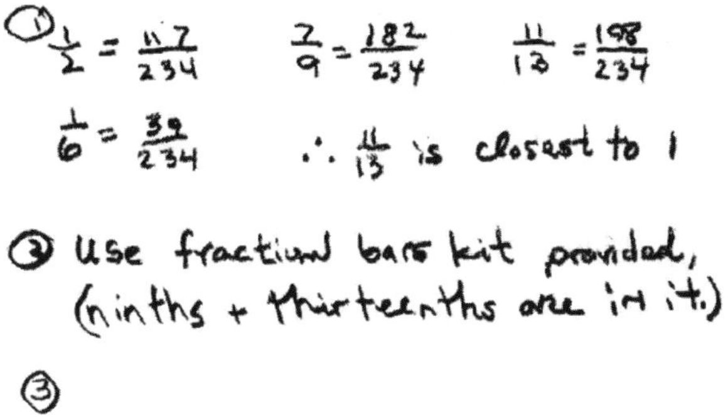

$$① \ \frac{1}{2} = \frac{117}{234} \qquad \frac{7}{9} = \frac{182}{234} \qquad \frac{11}{13} = \frac{198}{234}$$

$$\frac{1}{6} = \frac{39}{234} \qquad \therefore \frac{11}{13} \text{ is closest to } 1$$

② Use fraction bars kit provided,
(ninths + thirteenths are in it.)

③

 To analyze the post-assessment response associated with the pre-assessment response in Figure 5.13, go to question 6 in Looking Back.

Chapter Summary

This chapter focused on developing a range of strategies to order and compare different classes of fractions:

- reasoning with unit fractions (e.g., $\frac{1}{4} < \frac{1}{3}$ because fourths are smaller than thirds);
- extension of unit fraction reasoning to non-unit fractions (e.g., $\frac{7}{8} > \frac{4}{5}$ because $\frac{1}{8}$ away from a whole is smaller than $\frac{1}{5}$ away from a whole);
- reasoning based on models;
- reasoning through the use of a common reference fraction such as $\frac{1}{2}$; and
- reasoning involving equivalence.

Looking Back

1. Review Ted's response in Figure 5.14. While we cannot be sure, it is possible that Ted relied on a rule to compare $\frac{5}{6}$ to $\frac{3}{6}$. What are two reasoning strategies that Ted could have used to decide who planted more corn in their garden? Describe each.

Figure 5.14 Ted's response

Sam and Don each have a garden. The gardens are the same size. $\frac{5}{6}$ of Don's garden is planted with corn. $\frac{3}{6}$ of Sam's garden is planted with corn. Who has more corn in his garden?

When the denominators are the same just look at the numerators. 5 is bigger than 3. That means that $\frac{5}{6}$ is bigger than $\frac{3}{6}$.

2. Review the Everyday Mathematics Study Link 7.9 (Figure 5.3), and answer questions (a)–(d).
    (a) Which fraction pairs or sets of fractions provide the opportunity to use benchmarks to compare them? Explain your choices.
    (b) Which fraction pairs or sets of fractions provide the opportunity to use unit fraction reasoning to compare them? Explain your choices.
    (c) Identify fraction pairs or sets of fractions that would be difficult to compare, using models. Explain your choices.
    (d) Questions 7 and 8 ask students to explain how they compare the fraction pairs in questions 1 and 2 respectively. Provide an instructional/assessment rationale for students explaining questions 1 and 2.
3. Read through Mark's response to the problem in Figure 5.15.
    (a) Why did Mark's reasoning result in a correct solution to the problem?
    (b) Under what conditions would Mark's reasoning not work? Explain your answer with an example(s).
    (c) Provide a couple of examples of pairs of fractions you might ask Mark to compare to determine if he can extend his unit fraction

understanding to comparing other fractions. Provide a rationale
for each of the fraction pairs.

Figure 5.15 Mark's response

Linda hiked $\frac{1}{4}$ of the way up Mt. Mansfield. Jen hiked $\frac{1}{3}$ of the way up Mt. Mansfield.
Who hiked the farthest? Explain your answer using words and diagrams.

Jen hiked more because $\frac{1}{3}$
is bigger than $\frac{1}{4}$. If you were
going to split something in 3
parts it would be less than $\frac{1}{4}$
because the more you split something
the smaller the space gets.

4. Read through Tom's response to the problem in Figure 5.16.
   (a) What misunderstanding led Tom to conclude that both $\frac{3}{4}$ and $\frac{2}{3}$
       are closest to 1?
   (b) What additional questions might help Tom to understand why $\frac{3}{4}$
       and $\frac{2}{3}$ are not the same distance from 1 even though they are both
       "1 away" from a whole?

Figure 5.16 Tom's response

Which fraction is closest to 1?

$$\frac{1}{2} \quad \left(\frac{3}{4}\right) \quad \left(\frac{2}{3}\right) \quad \frac{1}{5}$$

Both $\frac{3}{4}$ and $\frac{2}{3}$ are closest to
1 because they are 1 away from
being whole.

5. Read through Kim's and Bob's responses to the same problem in
   Figures 5.17 and 5.18.
   (a) How did Kim and Bob use their knowledge of comparing proper
       fractions when they compared a mixed number to an improper
       fraction? Explain.

Figure 5.17 Kim's response

Susan ate $1\frac{1}{2}$ cupcakes and Billy ate $\frac{9}{8}$ cupcakes. Who ate more? Use words or diagrams to explain your answer.

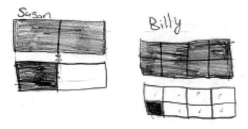

Figure 5.18 Bob's response

Susan ate $1\frac{1}{2}$ cupcakes and Billy ate $\frac{9}{8}$ cupcakes. Who ate more? Use words or diagrams to explain your answer.

> Susan ate more because $1\frac{1}{2}$ is bigger than $\frac{9}{8}$. I think $1\frac{1}{2}$ is bigger than $\frac{9}{8}$ because if you have $\frac{9}{8}$ it only goes over 1 by $\frac{1}{8}$. If you have $1\frac{1}{2}$ $\frac{1}{2}$ goes over one by $\frac{1}{2}$.

(b) Identify some mixed numbers/improper fractions that can be compared using benchmark reasoning to halves. Explain your choices.

(c) Identify some mixed numbers/improper fractions that can be compared using unit fraction understanding. Explain your choices.

6. Figure 5.19 below is the post-assessment response associated with the pre-assessment response shown in Figure 5.13. Identify the strategies used in Figure 5.19. Describe the evidence.

Figure 5.19 Teacher post-assessment response

Which fraction is closest to 1: $\frac{1}{2}$, $\frac{7}{9}$, $\frac{11}{13}$, or $\frac{1}{6}$?

Solve this problem using three different strategies.

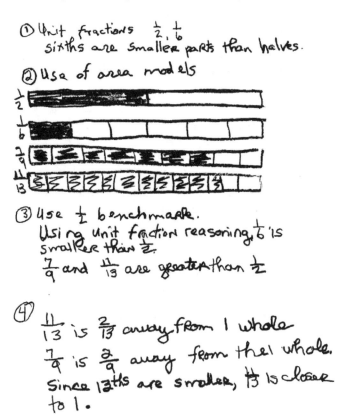

① Unit fractions $\frac{1}{2}$, $\frac{1}{6}$
   sixths are smaller parts than halves.

② Use of area models

   $\frac{1}{2}$

   $\frac{1}{6}$

   $\frac{7}{9}$

   $\frac{11}{13}$

③ Use $\frac{1}{2}$ benchmark.
   Using unit fraction reasoning, $\frac{1}{6}$ is
   smaller than $\frac{1}{2}$.
   $\frac{7}{9}$ and $\frac{11}{13}$ are greater than $\frac{1}{2}$

④ $\frac{11}{13}$ is $\frac{2}{13}$ away from 1 whole
   $\frac{7}{9}$ is $\frac{2}{9}$ away from the 1 whole.
   Since 13ths are smaller, $\frac{11}{13}$ is closer
   to 1.

Instructional Link—Your Turn

Use the guiding questions in Table 5.1 to help you think about your instruction and your math program: how do they support students as the students are learning to compare and order fractions?

Table 5.1 Instructional Link—strategies to support development of concepts related to comparing and ordering fractions.

Do you or does your program:	Yes/no
(1) provide opportunities for your students to compare fractions with different numerator and denominator combinations?	
(2) use models to compare and order fractions?	
(3) use models to develop and generalize reasoning strategies for comparing and ordering fractions?	
(4) encourage your students to use a variety of reasoning strategies when comparing or ordering fractions?	
(5) encourage and build on foundational skills to develop conceptual understanding and procedural fluency?	

Based on your analysis above, identify gaps in your instruction or mathematics program. How might you address these gaps?

Research Review—Comparing and Ordering Fractions

Researchers indicate that students should be able to compare and order fractions which contain different combinations of numerators and denominators (classes of fractions) (Behr, Wachsmuth, Post, & Lesh, 1984):

- *fractions with different numerators, but same denominators* (e.g., for $\frac{3}{6}$ and $\frac{5}{6}$, $\frac{3}{6} < \frac{5}{6}$);
- *fractions with the same numerators, but different denominators* (e.g., for $\frac{3}{8}$ and $\frac{3}{5}$, $\frac{3}{8} < \frac{3}{5}$);
- *fractions with different numerators and different denominators* (e.g., for $\frac{3}{8}$ and $\frac{5}{6}$, $\frac{3}{8} < \frac{5}{6}$).

Researchers have found that students use five types of reasoning when they **successfully** *compare and order fractions. Each type involves determining the relative contribution of the numerator and denominator to the overall size of the fraction* (Behr & Post, 1992):

- *reasoning with unit fractions (e.g., $\frac{1}{4} < \frac{1}{3}$ because fourths are smaller than thirds);*
- *extension of unit fractions reasoning to non-unit fractions (e.g., $\frac{7}{8} > \frac{4}{5}$ because $\frac{1}{8}$ (distance $\frac{7}{8}$ is away from a whole) is less than $\frac{1}{5}$ (distance $\frac{4}{5}$ is away from a whole));*
- *reasoning based on models* (VMP OGAP, student work samples, 2005);
- *reasoning through the use of a common reference fraction such as $\frac{1}{2}$; and*
- *reasoning involving equivalence.*

Behr & Post indicate that *"A child's understanding of the ordering of two fractions (that is, deciding which of the relations is equal to, is less than, or is greater than holds for two fractions) need to be based on an understanding of the ordering of unit fractions"* (1992, p. 21).

They found that *"Ultimately the problem of ordering two general fractions (different numerators and denominators) rests on considerable knowledge of fraction equivalence"* (Ibid., p. 23).

The evidence in the 2005 OGAP Exploratory Study indicated that the major strategy that students in grades 3–5 applied in the OGAP Student Post-Assessment to compare and order fractions involved the use of models. While the use of models to compare and order fractions appeared to result in a decrease in use of whole number reasoning found in the pre-assessment, developers saw few examples of students using unit fraction reasoning, extended unit fraction reasoning, equivalence ideas, or benchmarks (VMP OGAP, student work samples, 2005).

Two thirds of the pre-assessments sampled *(n = 67)* showed evidence of using only a model or common denominators in the pre-assessment (VMP OGAP (2006). [OGAP scale-up]. Unpublished raw data.).

6
Number Lines and Fractions

Big Ideas

Number lines can be used to help students to build an understanding of:

- the relative magnitude of fractions;
- equivalence;
- addition and subtraction of fractions; and
- the density of rational numbers.

Researchers *suggest that number lines can help to build an understanding of the magnitude of fractions, and be used to build concepts of equivalence and the density of rational numbers* (Behr & Post, 1992; Saxe et al., 2007).

Teachers in the OGAP studies found this as well. In particular, teachers found that using number lines helped students to think about a fraction as a number—allowing them to order, compare, and find equivalent fractions and to move away from using whole number reasoning as they worked with fractions (VMP OGAP, personal communication, 2005, 2006, 2007).

This chapter describes the characteristics of a number line that distinguish it from the area and set models. The chapter also contains descriptions of difficulties that students encounter when using a number line as well as instructional strategies that have been effective in helping students use a number line to develop an understanding of fraction concepts.

Characteristics of the Number Line

There are important characteristics that distinguish the number line from other fraction models (Bright, Behr, Post, & Wachsmuth, 1988).

1. *The unit is represented by a length as opposed to an area or a set of objects.*
2. *A number line requires symbols to define the unit, while the unit in an area or set of objects is implied in the model.*
3. *There are no visual separations between iterations of the units; that is,*

> *the units are continuous, unlike an area or set model in which the units are physically divided.*
> 4. *Units on number lines can be subdivided without restrictions.*

Figure 6.1 provides an example of how lengths are identified using points on a number line.

Figure 6.1 Number lines indicate lengths—on a number line a defined length represents the unit. A point on the line identifies a distance or length from 0. The point at 3 represents a length or distance of three units from 0

A length represents the unit

Figure 6.2 illustrates how the location of a fraction on a number line is dependent upon the symbols that define the unit.

Figure 6.2 Symbols define the unit—the accurate location of another number (e.g., $\frac{1}{2}$) on a number line is dependent on the symbols that define the unit. In contrast, no symbols are required to communicate $\frac{1}{2}$ in either the set or area models

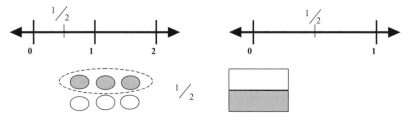

Figure 6.3 contrasts the continuous iteration of the units on a number line to the area models.

Figure 6.4 illustrates the unrestricted subdivision of units.

Difficulties that Students Encounter when Using Number Lines

Teachers reported *that when students first interacted with number lines they often reverted to their whole number reasoning* (Figure 6.5) *and placed fractions on the number line in order of the magnitude of their numerators or denominator* (VMP OGAP, personal communication, 2005, 2006, 2007).

This observation was confirmed with data from the OGAP 2005 Student Work Sub-study which found about 59 percent $\left(\frac{23}{39}\right)$ *of the student pre-assessment responses in the sample placed the fractions $\frac{1}{4}$ and $\frac{1}{3}$ as Ken did in his solution*

in Figure 6.5 (VMP OGAP, 2005. [Grade 4 pre-assessment question number 3]. Unpublished raw data.).

Figure 6.3 Continuous model—there are no visual separations between iterations of the units on a number line; that is, the units are continuous. In contrast, the units are physically separate in an area or set model

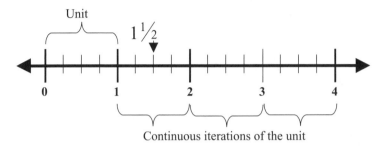

Continuous iterations of the unit

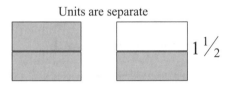

Figure 6.4 Subdivisions—the units on a number line can be subdivided into equal sub-units without restriction (i.e., to fourths, eighths, sixteenths, . . .). Pictured is a number line consisting of four units each subdivided into fourths

Subdivisions of the unit

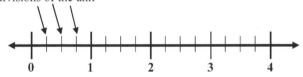

While area models can be subdivided without restriction, the units are separate; two wholes each subdivided into fourths.

When students first encounter number lines with multiple units it is not uncommon for them to find the fractional part of the whole line instead of locating the fraction relative to the defined unit (Mitchell & Horne, 2008; VMP OGAP, student work samples, 2005). Peter's response in Figure 6.6 exemplifies this error.

Figure 6.5 Ken's response—Ken used area models (circles) to represent $\frac{1}{2}$, $\frac{1}{4}$, and $\frac{1}{3}$. However, when Ken placed $\frac{1}{2}$, $\frac{1}{4}$, and $\frac{1}{3}$ on the number line, he incorrectly ordered them by the magnitude of the denominators

Place $\frac{1}{3}$ and $\frac{1}{4}$ on the number line.

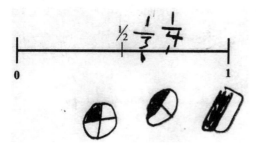

Figure 6.6 Peter's response—Peter found $\frac{1}{3}$ *of the line* using his part-to-whole understanding and ignoring the defined units (−1 to 0 and 0 to 1)

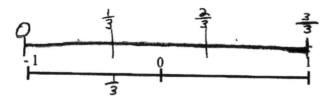

These errors may, in part, be explained with difficulties that researchers identified when students use number lines.

- *Students have difficulty integrating the visual model (line) and the symbols necessary to define the unit. The symbols and the tick marks that define the units and sub-units can act as distractors* (Behr, Lesh, Post, & Silver, as cited in Bright, Behr, Post, & Wachsmuth, 1988).
- *Students have a difficult time locating fractions on number lines that have been marked to show multiples of the unit or show marks to span from negative numbers to positive numbers* (Novillis-Larson, as cited in Behr & Post, 1992).
- *Students don't always understand that the numbers associated with points on a number line tell how far the points are from 0* (Pettito, 1990). For example, the two points marked 3 and −3 on a number line are both 3 units from 0.
- Researchers also "*hypothesize as long as partitioning and unpartitioning are difficult for children, number line representations of fractions may not be easily taught*" (Bright, Behr, Post, & Wachsmuth, 1988, p. 17).

Before discussing the research cited above, it is important to realize that "*students in the first three grades shift from sequential to proportional strategies to place numbers on a number line*" (Pettito, 1990, p. 57). This suggests that teachers should be sure that students are thinking proportionally when using a number line with whole numbers, not just sequentially, before asking students to locate fractions on a number line. See Figures 6.7 and 6.8.

Figure 6.7 Sequential thinking—whole numbers are placed on the number line considering only sequence, not the proportional distance between units that can define a length. The points at 2, 3, and 4 represent the second, third, and fourth numbers in the sequence, not the distance of 2, 3, and 4 equal units based on a defined unit 0 to 1

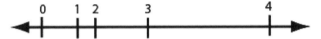

Any *number* can be placed *proportionally* on a number line once the unit (the distance from 0 to 1) is established (note: the distance between any two numbers could serve just as well to establish the unit). The numbers on the number line in Figure 6.8 are spaced proportionally; that is, 2 is twice as far from 0 as 1, 3 is three times as far from 0 as 1, and 4 is four times as far from 0 as 1. In terms of locating fractions on a number line proportionally, once the unit is established, any fraction can be placed on the number line. For example, the number $\frac{1}{2}$ is half as far from 0 as the number 1, $2\frac{1}{4}$ is $2\frac{1}{4}$ as far from 0 as 1, and so forth.

Figure 6.8 Proportional thinking—whole numbers are placed on the number line proportionally (equal intervals). The point at 4 represents a length or distance of four equal units from 0

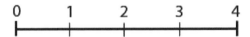

Students who are thinking sequentially, not proportionally may place fractions on a number line as Judy did in the response in Figure 6.9.

The location of $\frac{1}{4}$ suggests that Judi may not have progressed beyond sequential thinking. The mark showing her location of $\frac{1}{4}$ is not $\frac{1}{4}$ of the distance from 0 to 1. More evidence that Judi is thinking sequentially is Judi's justification of the locations of her fractions—that they belong where she placed them "because $\frac{1}{3}$ is bigger than $\frac{1}{4}$." While her comment is correct, it suggests that the magnitude of those two fractions is her only consideration in locating them on the number line.

Figure 6.9 Sequential response—Judy's response—The placement of $\frac{1}{4}$ and $\frac{1}{3}$ on this number line is sequential, not proportional. It appears that the student understands that $\frac{1}{2} > \frac{1}{3} > \frac{1}{4}$, but she did not place the fractions on the number line proportionally

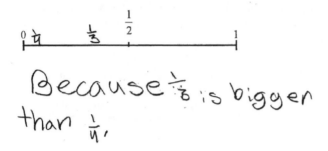

Students who are thinking proportionally will show evidence of partitioning proportionally in regards to the defined units on the number line. See Figures 6.10 and 6.11.

Figure 6.10 Proportional thinking—area models equal to the length of the number line were partitioned proportionally and then used to locate the fractions on the number line

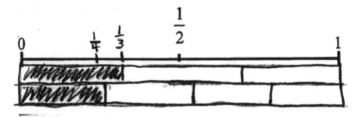

Figure 6.11 Proportional thinking—five-sixths is placed on the number line proportional to the distance of 0 to 1; that is, $\frac{5}{6}$ is about $\frac{5}{6}$ as far from 0 as the number 1

Locate $\frac{5}{6}$ on the number line below.

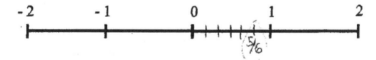

Introducing Number Lines into Instruction

"Although the number line is introduced to students in elementary school textbooks, its potential for students' learning has not been exploited by educators or researchers" (Saxe et al., 2007, p. 1). Teachers in OGAP studies began to embed the number line into instruction beyond what was provided in

their text materials as suggested by Saxe et al. (2007). (VMP OGAP, participant mini-projects, 2007).

First, research implies that strategies for engaging students in number lines may vary across grades. For example, when first introducing number lines involving fractions to young students, the research suggests that teachers should ensure that students are thinking about number lines with whole numbers proportionally, not sequentially.

Second, when moving from number lines with only whole numbers to using number lines to locate fractions, *researchers suggest that teachers use number lines with full knowledge of the difficulties that students may encounter* (Behr & Post, 1992). *Some teachers found that engaging students intentionally in the features of a number line that may later cause students difficulty made the use of the number line a more valuable instructional tool* (VMP OGAP, personal communication, 2007).

For example, teachers might use the number line in Figure 6.12 along with questions such as these to guide a discussion on the features of a number line.

1. Make a list of everything you notice about the number line. (The teacher uses the lists to guide a whole-class discussion.)
2. Identify where the number 4 is on this number line. What defined where the number 4 is located? What whole numbers are represented on this number line?
3. What do the tick marks between the numbers 1 and 2 indicate?
4. What numbers are represented on the number line for the tick marks to the left of 1? Are there other numbers between the tick marks on the number line? How could you determine what those numbers are? (See Chapter 7: Density of Fractions.)

Figure 6.12 Sample number line to use with guided questions

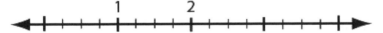

Another strategy adopted by some teachers and implied in the research is to vary the number lines presented to students. Number lines might contain single or multiple units, span from negative to positive numbers, or require partitioning or repartitioning. Figures 6.13 to 6.15 provide some examples of students who successfully located fractions on number lines with different structures.

In the problem in Figure 6.14, Kim was given a number line partitioned into sixths and was asked to locate the fractions $\frac{5}{12}$ and $\frac{3}{4}$ on the number line.

To solve the problem, Kim had to repartition the number line into twelfths and then recognize that $\frac{9}{12}$ was equivalent to $\frac{3}{4}$.

Figure 6.13 Asher's response—one-third is placed in the correct location on the number line that spans −1 to 1, providing evidence that the student integrated the visual model with the symbols

Place $\frac{1}{3}$ in the correct location on the number line below.

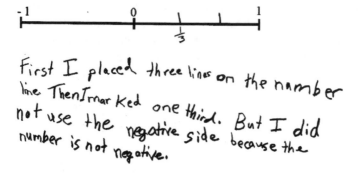

First I placed three lines on the number line. Then I marked one third. But I did not use the negative side because the number is not negative.

Figure 6.14 Kim's response—the student repartitioned the number line into twelfths and marked the location of $\frac{5}{12}$. Recognizing the equivalence of $\frac{3}{4}$ with $\frac{9}{12}$ allowed the student to accurately locate $\frac{3}{4}$ on the number line

(a) Place $\frac{5}{12}$ and $\frac{3}{4}$ in the correct location on this number line.

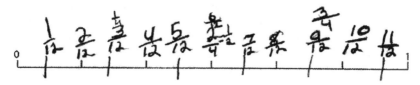

In Figure 6.15, Adam was given a number line that spans from −1 to +3 and used partitioning to correctly locate $\frac{8}{12}, \frac{8}{3}$, and $\frac{2}{3}$ on the number line.

Figure 6.15 Adam's response—to correctly locate $\frac{8}{12}, \frac{8}{3}$, and $\frac{2}{3}$ on the number line, Adam partitioned from 0 to 3 into thirds and from 0 to 1 into twelfths

Place $\frac{8}{12}, \frac{8}{3}$, and $\frac{2}{3}$ on the number line below in the correct position.

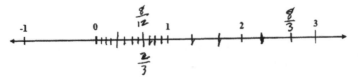

The number line in Figure 6.16 was used in a study conducted by Saxe et al. (2007). The problem includes a complete partitioning of the unit into fourths and also includes an incomplete partitioning of the unit into eighths. *Saxe et al. (2007) found that this number line with its missing partitions created challenges for students who lacked a strong conceptual sense of the magnitude of a fraction or who thought sequentially instead of proportionally.* For these reasons, number lines, like this one, which force students to use their proportional sense of the distance from 0 to the tic mark with the arrow in relation to the defined unit 0 to 1 can be a valuable instructional or assessment task.

Figure 6.16 A non-routine problem with incomplete partitioning into eighths (Saxe et al., 2007)

Name the fraction on this number line.

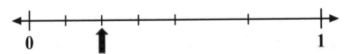

Evidence shows that translating the linear feature of part-to-whole models to number lines may help to explain improvement in the use of number lines (Bright, Behr, Post, & Wachsmuth, 1988; VMP OGAP, OGAP student work samples, 2005). This was a strategy adopted by many students and then capitalized on by teachers as evidenced in Laura's and Marko's responses found in Figures 6.17 and 6.18 (VMP OGAP, student work samples, 2005).

Figure 6.17 Laura's response—Laura drew area models under the number line and then used her partitioning of the linear feature of the area models to correctly locate the fractions on the number line

Place $\frac{8}{12}$, $\frac{8}{3}$, and $\frac{2}{3}$ on the number line below in the correct position.

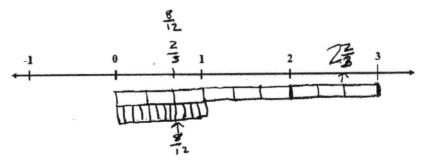

Figure 6.18 Marko's response—Marko partitioned an area model equal to one unit on the number line and then located $\frac{7}{8}$ relative to the defined unit

Place $\frac{7}{8}$ on the number line below.

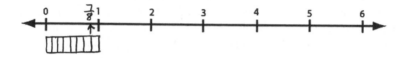

In order to reinforce the underlying structure of a number line and to strengthen understanding of the magnitude of fractions, *some teachers used a classroom-sized number line. The design allowed teachers from day to day to change the focus from one aspect of the number line to another (e.g., change the size of the unit or the number of units)* (VMP OGAP, personal communication, 2005, 2006, 2007). (See Figure 6.19.)

Figure 6.19 Two different classroom-sized number lines are displayed on a blackboard. Each number line has a different sized unit and a different number of units

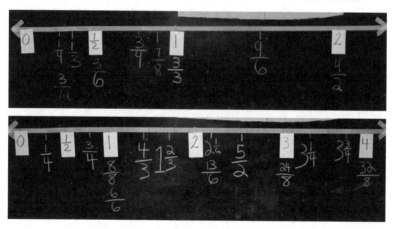

Student Use of Number Lines to Solve Problems

Many OGAP teachers encouraged students to use number lines to solve problems. *A preliminary analysis of 39 fourth-grade OGAP pre-assessments* $\left(\frac{39}{229}\right)$ *illustrates this point. Some 41 percent* $\left(\frac{16}{39}\right)$ *of the students used number lines to solve problems in their post-assessment responses. In contrast, only 7.7 percent* $\left(\frac{3}{39}\right)$ *of the students used the number line to solve problems in the pre-assessment* (VMP OGAP (2005). [Pre/post-assessment data]. Unpublished raw data.). (See Table 6.1.)

Examples of some ways that students used number lines to solve fraction problems are found in Figures 6.20 to 6.22. In Figure 6.20 Kaitlyn used a number line to solve a problem involving the density of fractions.

Table 6.1 Grade 4 OGAP—use of number lines pre-to post-assessment (VMP OGAP (2005). [Grade 4 pre-assessment]. Unpublished raw data.)

	Use of number lines	
	Pre-assessment	Post-assessment
Percent of students ($n = 39$)	7.7	41
Number of responses with number lines used to help solve problems*	3	43

* These data include only problems in which students use of a model is optional.

 See Chapter 7: Density of Fractions

Figure 6.20 Kaitlyn's response—Kaitlyn created a number line that she partitioned into twelfths and then used the partitions to identify fractions between $\frac{1}{3}$ and $\frac{5}{6}$

Name two fractions that are between $\frac{1}{3}$ and $\frac{5}{6}$.

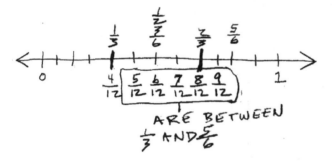

In Figure 6.21 Juan effectively used a number line to find the difference between the sum of $\frac{2}{3} + \frac{1}{4}$ and 1. He partitioned the number line into twelfths. He then used the partitioned number line to: (a) identify the distance from 0 to $\frac{1}{4}$ on the number line; (b) represent the addition of $\frac{1}{4}$ (using the distance identified) and $\frac{2}{3}$ by placing a segment equivalent to the length of $\frac{1}{4}$ at $\frac{2}{3}$ and, (c) identify the difference left as $\frac{1}{12}$ (circled on his solution).

Mathew (Figure 6.22) used a number line (and an area model) to compare two fractions.

Math Programs and Number Lines

To capitalize on the potential power of using the number line, teachers found themselves supplementing their mathematics program (VMP OGAP, personal conversations, 2005–2008). This is consistent with Saxe et al.'s observations.

We argue that number lines can support students' understanding of important properties of fractions. Fifth- and sixth-grade students can use the number line as a vehicle for understanding ideas like numerical units, relations between whole numbers and fractions, the density of the rational numbers (there are infinitely many rational numbers between any two), and although every number is unique, the number can be named in infinitely many ways (equivalence).

(2007, p. 1)

Figure 6.21 Juan's response—Juan effectively used a number line to solve a problem involving addition and subtraction of fractions

Tina ate $\frac{2}{3}$ of her candy and gave $\frac{1}{4}$ of her candy to her sister. She saved the rest of her candy. What is the fractional part of the candy that Tina saved?

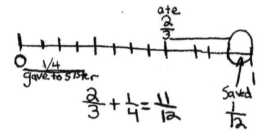

Figure 6.22 Mathew's response—Mathew effectively used a number line (and area model) to compare $\frac{2}{5}$ and $\frac{3}{10}$

There are some candies in a dish.

$\frac{2}{5}$ of the candies are chocolate.
$\frac{3}{10}$ of the candies are peppermint.

Are there more chocolate candies or more peppermint candies in the dish?

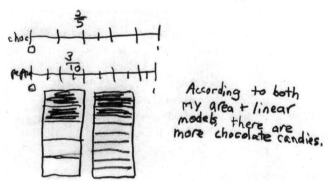

To find out how your program uses number lines in instruction, complete the Instructional Link activity (Table 6.2).

Measurement—A Direct Application of the Number Line

Some teachers have told us how surprised they are after completing a fraction unit to find that students had difficulty measuring with rulers. It appears that the students had not made the connection between the fraction concepts they had learned and their application when measuring with a ruler.

Bright, Behr, Post, & Wachsmuth indicated that *"the number line can be treated as a ruler"* (1988, p. 1). This appears to be a "which comes first" question —understanding of the number line and application to measurement or measuring with an application to the number line. While one can argue for a particular order over another, one teacher who used the number line first experienced huge payoffs.

> My students were surprised to learn that the tic marks on a ruler seemed to be related in size to their value just like a number line. The tic mark for $\frac{1}{2}$ was half of the distance from 0 to 1. The tic mark for $\frac{1}{4}$ was quarter the distance from 0 to 1. The tic mark for $1\frac{1}{2}$ is $1\frac{1}{2}$ times the distance from 0 to 1.
>
> (VMP OGAP, mini-project, 2007)

In any case, experience has shown that students have difficulty both measuring and using the number lines. Both, however, have the same conceptual foundation. To help think through the concept related to number lines and measurement, go to question 1 in Looking Back below.

Chapter Summary

This chapter presented ideas related to the use of number lines that research has shown help students as they develop their understanding of fraction concepts. In particular, the chapter focused on:

- the potential instructional benefits of using number lines as students develop an understanding of fraction concepts;
- the difficulties that students may encounter as they begin using number lines that should be attended to in instruction;
- instructional strategies using number lines;
- examples of how students use number lines to solve problems; and
- the importance of reviewing your mathematics program to assure that both the material and your instruction maximize the power of the number line to aid students as they develop their understanding of fractions.

 Looking Back

1. Measurement—a direct application of the number line

To help explore the relationship between measurement and number lines, respond to (a)–(c) below.

(a) Provide at least three important properties that number lines and measurement tools (such as rulers) share that have the potential to facilitate students' understanding of the connections between rulers and number lines.

(b) What are two important differences between number lines and scales on measuring tools?

(c) You provide your students with inch rulers, centimeter rulers, and strips of paper to measure. Their task is to measure each of the strips to the nearest eighth of an inch and tenth of a centimeter. Before your students begin using the rulers to measure the strips, identify three similarities and three differences between an inch ruler and a centimeter ruler that will allow students to measure strips accurately or identify challenges to accurate measurements. Use Figure 6.23.

Figure 6.23 Inch and centimeter rulers

What are three similarities and three differences between the inch ruler and a centimeter ruler below?

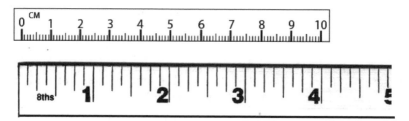

2. Mr. Brown had a large number line in the front of his classroom (Figure 6.24). On the first day that he used the number line he asked some students to place $\frac{1}{2}$ on the number where they thought it belonged. Mr. Brown had done no prior instruction with number lines, but thought this would be a good way to get information about what instructional issues he might face as students begin using number lines to solve fraction problems.

(a) The students were unsure where to locate $\frac{1}{2}$, but decided to locate it at the 3 on the number line. Is this correct or incorrect? Explain.

(b) What feature(s) of the number line may have been ignored by these students?

Figure 6.24 The number line in Mr. Brown's class

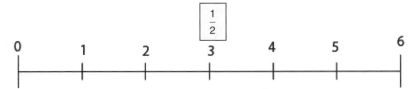

3. Look at Matt's response to the problem in Figure 6.25. What understandings and misunderstandings are evidenced in Matt's response?
Describe the evidence.

Figure 6.25 Matt's response

(Note: The original number line spanned from 0 to 1 and was partitioned into sixths.)

Place $\frac{5}{12}$ and $\frac{3}{4}$ on the number line below.

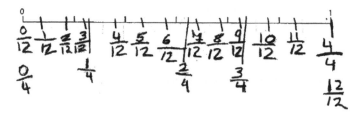

4. Nick (Figure 6.26) placed $\frac{1}{3}$ to the left of 0 on the number line.
   (a) What reasoning did Nick use to solve the problem? Describe
   the evidence.
   (b) What are some potential next instructional steps for Nick given
   the evidence in his work?

Figure 6.26 Nick's response

Place $\frac{1}{3}$ on the number line in the correct location.

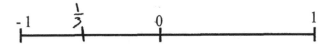

Instructional Link—Your Turn

Use the prompts in Table 6.2 to help you think about how your instruction and mathematics program provide students the opportunity to develop understandings of number lines.

Explain your answer using words or diagrams.

$\frac{1}{3}$ is not a whole number

Table 6.2 Instruction Link Strategies that support development of reasoning with fractions as quantities with an emphasis on instructional strategies that include number lines.

Do you or does your program:	*Yes/no*
(1) use number lines to build concepts of magnitude, equivalence and the density of rational number?	
(2) provide opportunities for students to engage regularly in problems involving number lines?	
(3) engage students in understanding the features of number lines?	
(4) provide opportunities for students to place fractions on number lines with more than one unit?	
(5) provide opportunities for students to place fractions on number lines with units of different sizes?	
(6) provide opportunities for students to place fractions on number lines that are already partitioned?	
(7) provide opportunities to solve non-routine problems involving number lines?	
(8) encourage students to use number lines to solve problems?	

Based on the analysis above, what gaps in your instruction or mathematics program did you identify? How might you address these gaps?

Research Review—Number Lines

Researchers *suggest that number lines can help to build understanding of the magnitude of fractions, and be used to build concepts of equivalence and the density of rational numbers* (Behr & Post, 1992; Saxe et al., 2007).

Teachers found that using number lines helped students to think about a fraction as a number—allowing them to order, compare, and find equivalent fractions. The number line was also helpful to move away from using whole number reasoning as students worked with fractions (VMP OGAP, personal communication, 2005, 2006, 2007).

There are important characteristics that distinguish the number line from other fraction models (Bright, Behr, Post, & Wachsmuth, 1988).

1. *The unit is represented by a length as opposed to an area or a set of objects.*
2. *A number line requires symbols to define the unit, while the unit in an area or set of objects is implied in the model.*
3. *There are no visual separations between iterations of the units; that is, the units are continuous, unlike an area or set model in which the units are physically divided.*
4. *Units of number lines can be subdivided without restrictions.*

Teachers reported *that when students first interacted with number lines, they often reverted to their whole number reasoning* (Figure 6.5) *and placed fractions on the number line in order of the magnitude of their numerators or denominators* (VMP OGAP, personal communication, 2005, 2006, 2007).

The OGAP 2005 Student Work Sub-study found that 58.9 percent $\left(\frac{23}{39}\right)$ *of the students in the sample placed the fractions* $\frac{1}{4}$ *and* $\frac{1}{3}$ *as Ken did in his solution in Figure 6.5 in their pre-assessment response* (VMP OGAP (2005). [Grade 4 pre-assessment question number 3]. Unpublished raw data.).

When students first encounter number lines with multiple units, it is not uncommon for them to find the fractional part of the whole line instead of locating the fraction relative to the defined unit (Mitchell & Horne, 2008; VMP OGAP, student work samples, 2005).

Students have difficulty integrating the visual model (line) and the symbols necessary to define the unit. The symbols and the tick marks that define the units and sub-units can act as distractors (Behr, Lesh, Post, & Silver, cited in Bright, Behr, Post, & Wachsmuth, 1988).

Students have a difficult time locating fractions on number lines that have been marked to show multiples of the unit or show marks to span from negative numbers to positive numbers (Novillis-Larson, cited in Behr & Post, 1992).

Students don't always understand that the numbers associated with points on a number line tell how far the points are from 0 (Pettito, 1990). For example, the two points marked 3 and −3 on a number line are both 3 units from 0.

Researchers also "*hypothesize as long as partitioning and unpartitioning are difficult for children, number line representations of fractions may not be easily taught*" (Bright, Behr, Post, & Wachsmuth, 1988, p. 17).

Researchers indicate that "*students in the first three grades shift from sequential to proportional strategies to place numbers on a number line*" (Pettito, 1990, p. 57).

"*Although the number line is introduced to students in elementary school textbooks, its potential for students' learning has not been exploited by educators or researchers*" (Saxe et al., 2007, p. 1).

Researchers suggest that teachers use number lines with full knowledge of the difficulties that students may encounter (Behr & Post, 1992).

Some teachers found that engaging students intentionally in the features of a number line that may later cause students difficulty made the use of the number line a more valuable instructional tool (VMP OGAP, personal communication, 2007).

Saxe et al. (2007) found that this number line with its missing partitions created challenges for students who lacked a strong conceptual sense of the magnitude of a fraction or who thought sequentially instead of proportionally.

Figure 6.16 A non-routine problem with incomplete partitioning into eighths (Saxe et al., 2007)

Name the fraction on this number line.

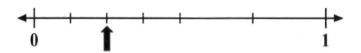

Evidence shows that translating the linear feature of part-to-whole models to number lines may help to explain improvement in the use of number lines (Bright, Behr, Post, & Wachsmuth, 1988; VMP OGAP, OGAP student work samples, 2005).

Some teachers used a classroom-sized number line. The design allowed teachers from day to day to change the focus from one aspect of the number line to another (e.g., change the size of the unit or the number of units) (VMP OGAP, personal communication, 2005, 2006, 2007).

A preliminary analysis of 39 fourth-grade OGAP pre-assessments $\left(\frac{39}{229}\right)$ illustrates this point; 41 percent $\left(\frac{16}{39}\right)$ of the students used number lines to solve problems in their post-assessment responses. In contrast, only 7.7 percent $\left(\frac{3}{39}\right)$ of the students used the number line to solve problems in the pre-assessment (VMP OGAP (2005). [Pre/post-assessment data]. Unpublished raw data.).

Table 6.1 Grade 4 OGAP—use of number lines pre-to post-assessment (VMP OGAP (2005). [Grade 4 pre-assessment]. Unpublished raw data.)

	Use of number lines	
	Pre-assessment	*Post-assessment*
Percent of students ($n = 39$)	7.7	41
Number of responses with number lines used to help solve problems*	3	43

* These data include only problems in which students use of a model is optional.

Teachers found themselves supplementing their mathematics program. This is consistent with Saxe et al.'s observations.

> *Although the number line is introduced to students in elementary school textbooks, its potential for students' learning has not been exploited by educators or researchers. We argue that number lines can support students' understanding of important properties of fractions. Fifth- and sixth-grade students can use the number line as a vehicle for understanding ideas like numerical unit, relations between whole numbers and fractions, the density of the rational numbers (there are infinitely many rational numbers between any two), and although every number is unique, the number can be named in infinitely many ways (equivalence).*

(2007, p. 1)

Bright, Behr, Post, & Wachsmuth (1988, p. 1) indicated that *"the number line can be treated as a ruler."*

7

The Density of Fractions

Big Ideas

- For any two given fractions, there is always another fraction between them.
- For any two given fractions, the number of fractions between them is infinite.
- Number lines can help students to understand the "betweenness" of fractions.

What is the Property of the Density of Fractions?

When studying fractions, students encounter a property of fractions that is very different from properties of whole numbers. It is the density property of fractions; that is, between any two fractions there are an infinite number of other fractions.

To understand this concept, let's investigate a TV remote control. A remote control for a television set has sound settings (volume) that show on the television screen as marked below. The remote works much like counting whole numbers.

In this example there are 10 settings for volume, from no volume to the loudest volume. (See Figures 7.1 and 7.2.)

Figure 7.1 The remote with 10 settings showing no volume

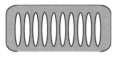

Figure 7.2 The remote set at the loudest setting

There is no compromise sound setting that is louder than setting 4 (Figure 7.3) but quieter than setting 5 (Figure 7.4). The whole numbers have this property; that is, there is no whole number between 4 and 5.

Figure 7.3 The remote at setting 4

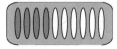

Figure 7.4 The remote at setting 5

Fractions have a very different property: between any two fractions, there are many different fractions. In fact, there are an infinite number of fractions between any two fractions. Students encounter this property in a number of situations when working with fractions. In mathematics class, for instance, students may be asked to name three fractions that are between $\frac{1}{4}$ and $\frac{3}{8}$.

However, this property is more than a mathematical exercise. It is encountered in everyday situations often without people realizing it. For example, unlike the remote control with a limited number of settings, imagine a volume control that can be placed at any position along a slide, like the example in Figure 7.5.

Figure 7.5 Volume control on a slide

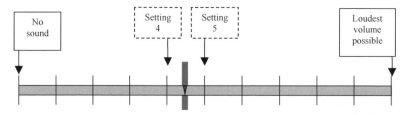

In this case, unlike the remote control, there are many positions between the fourth and fifth settings that are slightly louder than the fourth setting or softer than the fifth setting. For instance, the sound setting of $4\frac{1}{4}$ delivers a level that is louder than setting 4 and softer than level 5. The setting $4\frac{1}{7}$ provides a louder sound than 4, but a softer sound than $4\frac{1}{4}$. Every position on the slide can be approximated by a fraction $\left(e.g., 4\frac{30}{100}, 4\frac{31}{100}\right)$.

The volume control on a slide helps one begin to conceptualize the idea of density. However, to truly comprehend the density property, we can start with two different fractions such as $\frac{1}{4}$ and $\frac{1}{2}$, and see how to develop an infinite number of fractions that are between them. As a start, it should be clarified what it means to say "an infinite number of fractions between $\frac{1}{4}$ and $\frac{1}{2}$". It means that a list of all the different fractions between $\frac{1}{4}$ and $\frac{1}{2}$ has no end; that is, no matter how many fractions have been found, there are still more.

Density—The Mathematics

One way to mathematically demonstrate the *density property of fractions* is by looking at averages. The average of any two different numbers will always be another number between the two numbers (see Figure 7.6).

Figure 7.6 The average of two numbers results in another number between the two numbers. We will start with 60 and 100 and use averages to generate an endless list of numbers, all of them between 60 and 100

The average of 60 and 100 is 80 ($\dfrac{60 + 100}{2} = 80$).

The average of 80 and 100 is 90 ($\dfrac{80 + 100}{2} = 90$).

The average of 90 and 100 is 95 ($\dfrac{90 + 100}{2} = 95$).

The average of 95 and 100 is 97.5 ($\dfrac{95 + 100}{2} = 97.5$).

These averages give us the start of a list $60 < 80 < 90 < 95 < 97.5 < \ldots < 100$. (Note: "$\ldots < 100$" in the previous line indicates that the list we are generating is an endless list (an *infinite* list) of numbers that are between 60 and 100.)

The number line allows geometry to help us to understand the average of two numbers. For two different numbers, their average is located on the number line exactly midway between them. (See the location of 80 midway between 60 and 100 in Figure 7.7.) Starting at 80 and going halfway to 100 brings us to

Figure 7.7 The average of 60 and 100 is 80 or the midpoint between 60 and 80. The average of 80 and 100 is 90 or the midpoint between 80 and 90. The average of 90 and 100 is 95 or the midpoint between 90 and 100

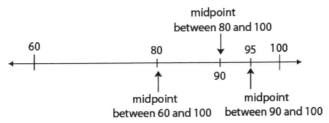

90, the average of 80 and 100. Next, the average of 90 and 100 is 95, and the average of 95 and 100 is 97.5 (which is halfway from 95 to 100). Notice when we start at the previous average and go halfway to 100, we can never reach 100. We have an endless list of averages, and these averages form an *infinite list* of numbers between 60 and 100. (Note: this averaging process will be referred to in the rest of the chapter as "successive averages.")

We will now continue with the idea of successive averages to investigate the infinite number of fractions between $\frac{1}{4}$ and $\frac{1}{2}$. Figure 7.8 illustrates how halving the distance between $\frac{1}{4}$ and $\frac{1}{2}$ (averaging of $\frac{1}{4}$ and $\frac{1}{2}$) results in the fraction $\left(\frac{3}{8}\right)$ that is exactly midway on the number line between $\frac{1}{4}$ and $\frac{1}{2}$. While successive averages $\left(\frac{3}{8}, \frac{5}{16}, \frac{9}{32}, \ldots > \frac{1}{4}\right)$ get closer and closer to $\frac{1}{4}$, the average will never reach $\frac{1}{4}$ because there will always be a fraction halfway between $\frac{1}{4}$ and the fraction being averaged.

The density of fractions property applies to mixed numbers as well. See Figure 7.9. By successive averaging starting with $1\frac{1}{4}$ and $1\frac{1}{2}$, we will find an infinite list of mixed numbers that are between $1\frac{1}{4}$ and $1\frac{1}{2}$, but that never reach $1\frac{1}{4}$ $\left(1\frac{3}{8} > 1\frac{5}{16} > 1\frac{9}{32}, \ldots > 1\frac{1}{4}\right)$.

The number lines in Figures 7.8 and 7.9 help us to understand an important idea about how density between proper fractions is related to density between mixed numbers. What we can see is that each fraction on the number line in

Figure 7.8 Three-eighths is the midpoint between $\frac{1}{4}$ and $\frac{1}{2}$. Five sixteenths is the midpoint between $\frac{3}{8}$ and $\frac{2}{8}$. The midpoint between $\frac{4}{16}$ and $\frac{5}{16}$ is $\frac{9}{32}$

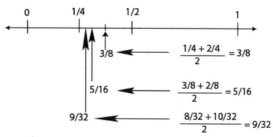

Figure 7.9 The fractions on the number line illustrate that the same rational numbers that can be mapped between 0 and 1 can be mapped onto the interval between 1 and 2

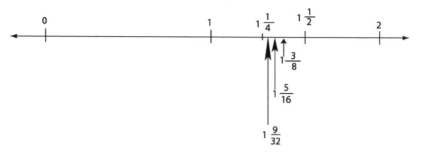

Figure 7.9 is exactly one more than a corresponding fraction on the number line in Figure 7.8. For the infinite list of fractions we generated that were between $\frac{1}{4}$ and $\frac{1}{2}$, adding 1 to each of these gives us an infinite list of mixed numbers that are between $1\frac{1}{4}$ and $1\frac{1}{2}$.

To accept as true that there are an infinite number of fractions between *any* two fractions requires integrating these ideas:

1. The successive averaging of fractions in Figure 7.8 (starting with averaging of $\frac{1}{4}$ and $\frac{1}{2}$) illustrates the use of averages to generate an unending list of different fractions, each average closer to $\frac{1}{4}$ than the previous average. Since all of these averages are between $\frac{1}{4}$ and $\frac{1}{2}$, we have generated an infinite list of fractions that are between $\frac{1}{4}$ and $\frac{1}{2}$.
2. While the example used focused on the fractions $\frac{1}{4}$ and $\frac{1}{2}$, successive averages between any two different fractions would have produced the same effect. For those two different fractions, there are infinitely many fractions between them.
3. Any fraction that is located between 0 and 1 can be mapped onto any other consecutive whole number interval as was explained above and modeled in Figure 7.9.
4. Since there are an infinite number of proper fractions between any two fractions and they can be mapped to other intervals, there is an infinite number of fractions between any two numbers.

The next section illustrates how the density of rational number property is applied in the design of measurement tools.

The Connection of "Betweenness" of Fractions to Accuracy in Measurements

Measurement is another common area in which people encounter this "betweenness" property. As you read the section, think about how the halving process (averaging) described above might be used to manufacture measurement instruments of varying accuracy. For example, one may use a ruler (or other scale) to measure something that requires greater accuracy than the tool allows. In Figure 7.10 the object being measured has a length between $1\frac{5}{8}$

Figure 7.10 The picture is about $1\frac{11}{16}$ inches long

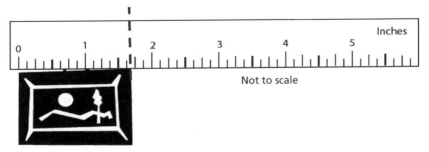

inches and $1\frac{3}{4}$ inches. To determine a more exact length of the object would require an understanding that there are fractions between $1\frac{5}{8}$ inches and $1\frac{3}{4}$ inches. In fact, a new ruler that measures in sixteenths of an inch could allow us to decide that the length is about $1\frac{11}{16}$ inch.

To connect this directly to the concept of density, each mark on the new ruler to sixteenths would be based on averaging adjacent numbers on the pictured ruler. For example, $1\frac{11}{16}$ is the average of $1\frac{5}{8}$ and $1\frac{3}{4}$.

While the example in Figure 7.10 may seem simplistic, the idea and its applications are not. The concept of the density of rational numbers is applied when making measuring tools designed for different purposes. For example, a carpenter building a barn would be content with accuracy to the nearest eighth of an inch. However, a cabinet maker would not. The cabinet maker would need accuracy to thirty-seconds of an inch to assure that components of the cabinet fit together. However, thirty-seconds of an inch would be wholly inadequate for someone making electronic components as electronic components need to be accurate to microns (forty millionth of an inch). Imagine the repartitioning necessary to go from a thirty-second of an inch to a forty millionth of an inch.

To engage students in the concept, you may want them to interview people who use measurement in their work or have students research the accuracy needed to produce different products such as:

an iPod
a bicycle
a car
an airplane
other.

Developing Students' Understanding of the Concept

In spite of the importance of this concept *researchers indicate that students have a difficult time understanding and applying the property of the density of rational numbers* (Orton et al., 1995).

This finding was supported in an assessment of prospective teachers' knowledge of rational numbers involving 147 first-year elementary majors. Tirosh, Fischbein, Graeber, & Wilson (1998) *found that "only 24 percent knew that there was an infinite amount of numbers between $\frac{1}{5}$ and $\frac{1}{4}$, 43 percent claimed that there are no numbers between $\frac{1}{5}$ and $\frac{1}{4}$, and 30 percent claimed that $\frac{1}{4}$ is the successor of $\frac{1}{5}$"* (pp. 8–9).

These findings are also supported with data from a VMP OGAP 2005 substudy based on two questions assessing fifth-grade students' understanding of the density of fractions in the OGAP pre- and post-assessments. The first question asked students to name two fractions that are between $\frac{1}{3}$ and $\frac{3}{4}$. The second asked the more general question about the concept—do you think that there are any other fractions besides the ones you identified that are between $\frac{1}{3}$ and $\frac{3}{4}$?

Only about 45.7 percent $\left(\frac{16}{35}\right)$ of the fifth-grade students in the sample (35 students/151 grade 5 students) *correctly identified two fractions between* $\frac{1}{3}$ and $\frac{3}{4}$ in the pre-assessment and 60 percent $\left(\frac{21}{35}\right)$ in the post-assessment (VMP OGAP (2005). [Grade 5 pre- and post-assessment question number 5]. Unpublished raw data.). See Table 7.1.

Table 7.1 Percentage of sample that identified no fractions, one fraction, or two fractions in the grade 5 OGAP pre- and post-assessments

	Percentage of sample ($n = 35$)	
Number of fractions identified	OGAP pre-assessment	OGAP post assessment
0	25.7	11.4
1	28.6	28.6
2	45.7	60.0

Richard's response (Figure 7.11) is typical of solutions in which only one fraction was named. While we don't know for sure, Richard may have used his benchmark understanding recognizing that $\frac{1}{3} < \frac{1}{2} < \frac{3}{4}$, but was unable to apply another strategy or understanding to naming another fraction between $\frac{1}{3}$ and $\frac{3}{4}$.

While Richard (Figure 7.11) seemed limited in his understanding, other students, like Madison (Figure 7.12), either misunderstood the problem or the concept. These students identified two equivalent fractions $\left(\text{e.g., } \frac{2}{3} \text{ and } \frac{4}{6}\right)$, not two different fractions.

Figure 7.11 Richard's response—which leads one to believe that he assumed that there were only two fractions between $\frac{1}{3}$ and $\frac{3}{4}$

Name two fractions that are between $\frac{1}{3}$ and $\frac{3}{4}$.

½

Don't Know the otherone.

Figure 7.12 Madison's response—Madison identified two equivalent fractions that are between $\frac{1}{3}$ and $\frac{3}{4}$

(a) Name two fractions that are between $\frac{1}{3}$ and $\frac{3}{4}$.

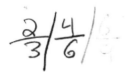

While Madison is correct that $\frac{2}{3}$ and its equivalent $\frac{4}{6}$ are between $\frac{1}{3}$ and $\frac{3}{4}$, they are not different fractions, just different names for the same fraction (see Chapter 8: Equivalent Traditions and Comparisons).

Researchers indicate that using number lines has the potential to help build an understanding of the density of rational number concept (Saxe et al., 2007).

Todd's (Figure 7.13) and Kaitlyn's (Figure 7.14) responses show some ways in which students use number lines to solve problems involving "betweenness." Todd partitioned a number line into twelfths. This enabled him to identify twelfths found between $\frac{1}{3}$ and $\frac{3}{4}$.

Kaitlyn (Figure 7.14) partitioned two number lines of equal length; one into fourths and the other into sixths. This enabled her to see the "betweenness" involved in the problem and identify two different fractions that are between $\frac{1}{3}$ and $\frac{3}{4}$.

Figure 7.13 Todd's response—Todd partitioned a number line into twelfths which enabled him to identify four fractions between $\frac{1}{3}$ and $\frac{3}{4}$

Name two fractions that are between $\frac{1}{3}$ and $\frac{3}{4}$.

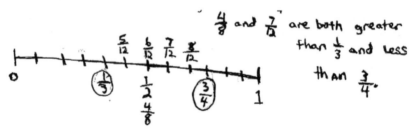

Figure 7.14 Kaitlyn's response—Kaitlyn partitioned two number lines and successfully identified $\frac{1}{2}$ and $\frac{4}{6}$ as fractions that are between $\frac{1}{3}$ and $\frac{3}{4}$

Name two fractions that are between $\frac{1}{3}$ and $\frac{3}{4}$.

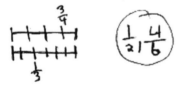

Students on the OGAP pre- and post-assessment were not successful with the second more general density of fractions question: "Do you think there are any other fractions besides the ones you identified that are between $\frac{1}{3}$ and $\frac{3}{4}$?" *Only one student response in the OGAP pre-assessment sample* $\left(\frac{1}{35}\right)$ *and two student responses* $\left(\frac{2}{35}\right)$ *in the post-assessment showed any evidence of understanding the infinite nature of the density of rational numbers* (VMP OGAP (2005). [Grade 5 pre- and post-assessment question number 5]. Unpublished raw data.).

Ava's response (Figure 7.15) illustrates a developing understanding that there are an infinite number of fractions between $\frac{1}{3}$ and $\frac{3}{4}$.

Figure 7.15 Ava's response—Ava recognized that there are an unlimited number of fractions between $\frac{1}{3}$ and $\frac{3}{4}$

Do you think that there are any other fractions besides the ones you identified that are between $\frac{1}{3}$ and $\frac{3}{4}$?

Yes, There are because there is an unlimited number of fractions between $\frac{1}{3}$ and $\frac{3}{4}$.

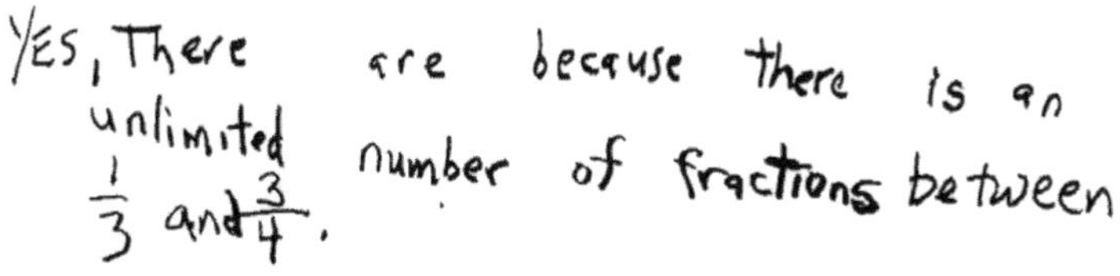

The most common (57 percent of the sample $\left(\frac{20}{35}\right)$) misconception evidenced for the more general question involved students explicitly stating or alluding to there being a finite set of fractions between the two fractions (VMP OGAP (2005). [Grade 5 pre- and post-assessment question number 5]. Unpublished raw data.). This type of response represents a developing understanding; that is, they are able to name other fractions, but have not generalized the concept to recognize that there are an infinite number of fractions between any two fractions.

Todd's response (Figure 7.16) to the more general question about the density of fractions is an example of this developing understanding. Todd explicitly states that "I only named 2 of 5 numbers" inferring that there is a finite set of fractions between the two fractions.

Figure 7.16 Todd's response—which suggests that he believes that there is a limited number of fractions between $\frac{4}{12}$ and $\frac{9}{12}$

Do you think that there are any other fractions besides the ones you identified that are between $\frac{1}{3}$ and $\frac{3}{4}$?

Yes, because $\frac{1}{3} = \frac{4}{12}$ and $\frac{9}{12} = \frac{3}{4}$ and there are 5 numbers in between. I only named 2 out of 5 numbers;

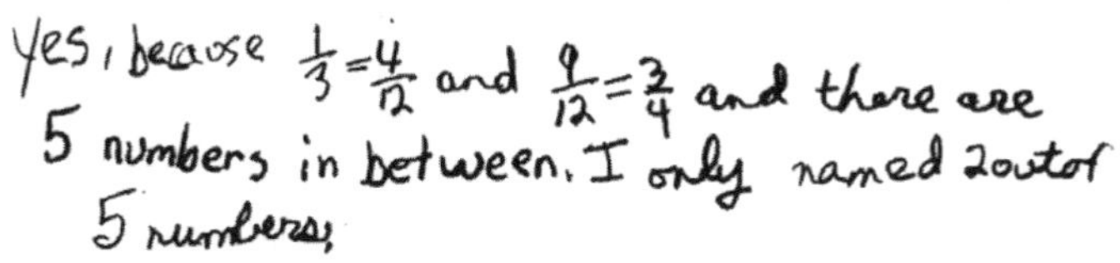

Renee's response in Figure 7.17 illustrates the second most common error found in the OGAP post-assessment sample related to the general question about the density of fractions. *Some 26 percent of the sample indicated that there are more fractions between $\frac{1}{3}$ and $\frac{3}{4}$, but identified equivalent fractions, not different fractions* (VMP OGAP (2005). [Grade 5 post-assessment question number 5]. Unpublished raw data.).

Because number lines provide a visual picture of the "betweenness" of fractions, they can be used to extend the developing understanding of each of these students: Richard (Figure 7.11) who named only one fraction; Madison (Figure 7.12) and Renee (Figure 7.17) who named equivalent fractions instead of different fractions; and Todd (Figures 7.13 and 7.16) who used a number line to identify some fractions, but who did not extend that to a more general understanding.

Figure 7.17 Renee's response—which describes finding equivalent fractions forever

Do you think that there are any other fractions besides the ones you identified that are between $\frac{1}{3}$ and $\frac{3}{4}$? Explain why or why not.

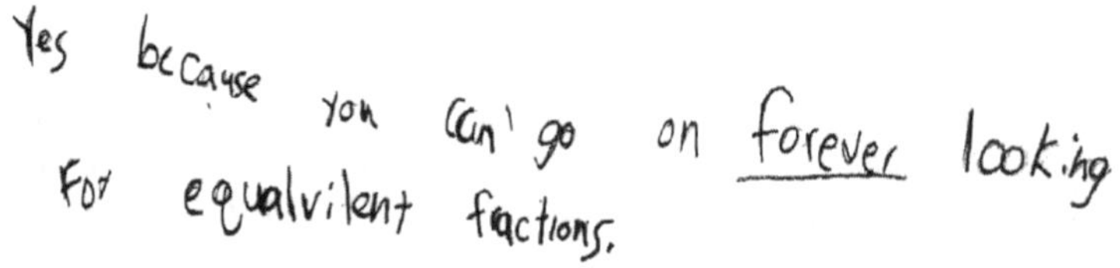

Yes because you can' go on forever looking
For equalvilent factions.

Question 2 in Looking Back on p. 128 provides an opportunity for you to think about how number lines can be used to extend each of these student's developing understanding of the density of fractions.

Density at the Elementary and Middle School

Most elementary and middle-school students' experience with density is limited to identifying fractions between fractions. Teachers should carefully select pairs of fractions that strategically and thoughtfully expand their students' ability to identify fractions between fraction pairs and to develop their understanding of the generalized concept.

For example, on the surface it might feel that one should first give students fraction pairs with common denominators, such as $\frac{5}{12}$ and $\frac{11}{12}$. However, it is possible that this strategy might lead students to an incorrect generalization that fractions can be counted like whole numbers. Kim's response in Figure 7.18 may be an example of a student generalizing her whole number understanding of counting numbers.

On the other hand, choosing a pair of fractions based on students' experience is important as well. For younger students, asking them to find two fractions between $\frac{1}{3}$ and $\frac{1}{4}$ may be unreasonable as the fractions are very close to each other.

You may want to start with fractions such as $\frac{1}{3}$ and $\frac{3}{4}$. These fractions are not so close together, are familiar fractions, and are on either side of the benchmark $\frac{1}{2}$. Over time, however, students should be able to use their understanding of partitioning and equivalence to identify fractions between a variety

Figure 7.18 Kim's response—Kim may have used her whole number understanding of counting numbers

Name three fractions that are between $\frac{5}{12}$ and $\frac{11}{12}$.

That's easy cause I
can count 6, 7, 8, 9, 10,

$$\frac{6}{12} \quad , \quad \frac{7}{12} \quad) \quad \frac{8}{12} \quad) \quad \frac{9}{12} \quad) \quad \frac{10}{12}$$

of given fractions. Older students should solve density of rational numbers problems involving fraction pairs that are very close, such as $\frac{1}{10}$ and $\frac{1}{11}$, and fraction pairs that include mixed numbers and improper fractions.

Finding fractions between fractions that are "very close" requires a more generalized understanding and greater flexibility than partitioning a number line as Todd did in Figure 7.13 or using benchmark fractions as Richard did in Figure 7.11.

 Question 3 in Looking Back on p. 128 provides an opportunity for you to determine fractions between pairs of fractions using a range of reasoning strategies.

Chapter Summary

This chapter focused on the concept of the density of fractions with an emphasis on:

- the concept of density;
- misunderstandings that students have as they are identifying fractions between fraction pairs and developing an understanding of the generalized concept; and
- the role that number lines can play in building an understanding of the "betweenness" of fractions.

Looking Back

1. Review Seth's response in Figure 7.19 and then answer the following questions.
   (a) How did Seth (Figure 7.19) use his understanding of partitioning to answer Part (a) of the question?
   (b) Based on Seth's response to Part (b), what are the strengths and limitations of his partitioning strategy?

Figure 7.19 Seth's response

(a) Name two fractions that are between $\frac{1}{3}$ and $\frac{3}{4}$.

(b) Do you think that there are any other fractions besides the ones you identified that are between $\frac{1}{3}$ and $\frac{3}{4}$?

There are Probably more. But I am not totally sure..

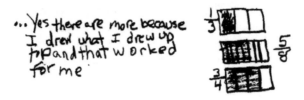

... Yes there are more because I drew what I drew up top and that worked for me

2. It was suggested in the chapter that number lines could be used to extend the developing understanding of Richard (Figure 7.11), Madison (Figure 7.12), and Todd (Figures 7.13 and 7.16). Review each response and then answer (a), (b), and (c).
   (a) Richard (Figure 7.11) named only one fraction. Provide an example of a way the number line could extend his understanding to identifying different fractions between $\frac{1}{3}$ and $\frac{3}{4}$ besides $\frac{1}{2}$.
   (b) Madison (Figure 7.12) named equivalent fractions instead of other fractions. Provide an example of a way the number line could extend her thinking beyond equivalent fractions.
   (c) Todd (Figures 7.13 and 7.16) used a number line to identify some fractions, but did not extend that to a more general understanding. Provide an example of a way the number line could extend his thinking beyond equivalent fractions.
3. Find three different fractions between the fraction pairs listed at (i), (ii), and (iii), using two different strategies for each fraction pair. Then answer questions (a), (b), and (c).
   (a) What difficulties do you think students might encounter as they solve these problems?
   (b) What kinds of error might result from these difficulties?
   (c) As a set of questions, what information can the student work provide that the evidence from a single question might not provide?
   (i) $\frac{4}{10}$ and $\frac{7}{10}$;

(ii) $\frac{1}{8}$ and $\frac{1}{4}$;

(iii) $\frac{1}{10}$ and $\frac{1}{9}$.

Instructional Link—Your Turn

Use the guiding questions in Table 7.2 to help you think about how your instruction or mathematics programs provide students with the opportunity to solve problems involving the density of rational numbers.

Ask yourself—is there anything in my **instruction or mathematics program** (activities, games, lessons, problems) that INTENTIONALLY provides opportunities for students to transfer their knowledge of partitioning and equivalence to identify a fraction or fractions that are between any two given fractions?

Table 7.2 Instructional Link strategies to support development of concepts related to the density of fractions

Do you or does your program:	Yes/no
(1) provide opportunities for students to use number lines to develop and expand their understanding of density?	
(2) make a connection between density of fractions and accuracy in measurements?	
(3) provide students with opportunities to solve problems that promote a clear conceptualization of the density of fractions?	

Based on the analysis above, what gaps in your instruction or mathematics program did you identify? How might you address these gaps?

Research Review—Density of Fractions

Researchers indicate that students have a difficult time understanding and applying the property of the density of rational numbers (Orten et al. 1995).

Tirosh, Fischbein, Graeber, & Wilson (1998) *found that "only 24 percent knew that there was an infinite amount of numbers between $\frac{1}{5}$ and $\frac{1}{4}$, 43 percent claimed that there are no numbers between $\frac{1}{5}$ and $\frac{1}{4}$, and 30 percent claimed that $\frac{1}{4}$ is the successor of $\frac{1}{5}$"* (pp. 8–9).

Only about 45.7 percent $\left(\frac{16}{35}\right)$ of the fifth-grade students in the sample (35 students/151 grade 5 students) *correctly identified two fractions between $\frac{1}{3}$ and $\frac{3}{4}$ in* the pre-assessment and 60 percent $\left(\frac{21}{35}\right)$ in the post-assessment (VMP OGAP (2005). [Grade 5 pre- and post-assessment question number 5]. Unpublished raw data.). See Table 7.1.

Researchers indicate that using number lines has the potential to help build an understanding of the density of rational number concept (Saxe et al., 2007).

Table 7.1 Percentage of sample that identified no fractions, one fraction, or two fractions in the grade 5 OGAP pre- and post-assessments

	Percentage of sample (n = 35)	
Number of fractions identified	*OGAP pre-assessment*	*OGAP post assessment*
0	25.7	11.4
1	28.6	28.6
2	45.7	60.0

Only one student response in the OGAP pre-assessment sample $\left(\frac{1}{35}\right)$ *and two student responses* $\left(\frac{2}{35}\right)$ *in the post-assessment showed any evidence of understand-ing the infinite nature of the density of rational numbers* (VMP OGAP (2005). [Grade 5 pre- and post-assessment question number 5]. Unpublished raw data.).

The most common $\left(57 \text{ percent} \left(\frac{20}{35}\right)\right)$ *error evidenced for the more general question involved students explicitly stating or alluding to there being a finite set of fractions between the two fractions* (VMP OGAP (2005). [Grade 5 pre- and post-assessment question number 5]. Unpublished raw data.).

Some 26 percent of the sample indicated that there are more fractions between $\frac{1}{3}$ *and* $\frac{3}{4}$, *but identified equivalent fractions, not different fractions* (VMP OGAP (2005). [Grade 5 post-assessment question number 5]. Unpublished raw data.).

8

Equivalent Fractions and Comparisons

Big Ideas

- Saying that two fractions are equivalent is saying that the two fractions are different names (symbols) for the same number.
- There are an infinite number of different names for a given fraction.
- Understanding equivalence and having an efficient procedure to find equivalent fractions are critical as students encounter problems involving comparing, ordering, and operating with fractions.

"Fraction order and equivalence ideas are fundamentally important concepts. They form the framework for understanding fractions and decimals as quantities that can be operated on in meaningful ways" (Post et al., 1993, p. 15). Understanding equivalence of fractions is crucial to a student's ability to add and subtract fractions and to compare and order fractions.

However, researchers say that *"students who really do not understand what a fraction means will have a hard time finding another fraction equivalent to it"* (Bezuk & Bieck, 1993, p. 129). In addition, some students *"have a continuing interference from their knowledge of whole numbers"* (Post, Behr, Lesh, & Wachsmuth, 1986).

Samir, Tania, and Corey (Figures 8.1, 8.2, and 8.3 respectively) show a range of understanding about equivalence, from Samir who is still experiencing "interference" from whole number reasoning, to Tania who uses a model to effectively compare the two fractions, to Corey who found equivalent fractions.

Tania's solution includes an effective use of area models representing $\frac{3}{8}$ and $\frac{1}{4}$. Her solution presents evidence that she is ready to engage in discussions about equivalent fractions and their use in solving problems.

Corey used an understanding he has developed about equivalent fractions to make a decision about whether the dish contains more chocolate or peppermint candies.

Figure 8.1 Samir's response—Samir incorrectly compared the magnitude of the denominators and numerators, not the magnitude of the fractions

There are some candies in a dish.

$\frac{2}{5}$ of the candies are chocolate.
$\frac{3}{10}$ of the candies are peppermint.

Are there more chocolate candies or more peppermint candies?

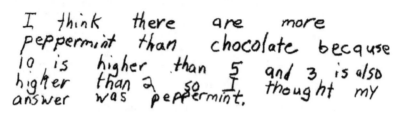

Figure 8.2 Tania's response—Tania used a model to compare $\frac{3}{8}$ and $\frac{1}{4}$

There are some candies in a dish.

$\frac{3}{8}$ of the candies are chocolate.
$\frac{1}{4}$ of the candies are peppermint.

Are there more chocolate candies or more peppermint candies?

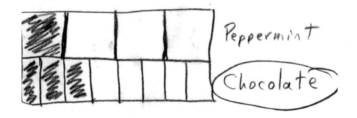

This chapter focuses on the concept of equivalence and on how students develop an understanding of equivalence.

Understanding the Concept

Two properties are central to the equivalence of fractions:

1. Saying that two fractions are equivalent is saying that the two fractions are both names (symbols) for the same number.
2. The number of different names for a given fraction is infinite. For example, the fraction $\frac{1}{2}$ can be expressed by any of the names (symbols) $\frac{1}{2} = \frac{2}{4} = \frac{3}{6} = \frac{4}{8} = \frac{5}{10} \ldots$, and so on.

One way to understand these properties is to think about the concept in terms of a number line. Because all equivalent fractions have the same value, they are located at the same point on a number line. The number line in

Figure 8.3 Corey's response—Corey used an understanding of equivalent fractions $\left(\frac{2}{5}=\frac{4}{10}\right.$ and $\left.\frac{3}{10}=\frac{1\frac{1}{2}}{5}\right)$ to compare the fractions $\frac{2}{5}$ and $\frac{3}{10}$

There are some candies in a dish.

$\frac{2}{5}$ of the candies are chocolate.
$\frac{3}{10}$ of the candies are peppermint.

Are there more chocolate candies or more peppermint candies?

Figure 8.4 is partitioned into sixteenths and can be used to illustrate that $\frac{3}{4} = \frac{6}{8} = \frac{12}{16}$. If one continued to partition the number line into thirty-seconds, sixty-fourths, one hundred and twenty-eights, . . . one can picture an infinite number of possible names for $\frac{3}{4}$, all stacked on top of the $\frac{12}{16}$ that is stacked on top of both $\frac{6}{8}$ and $\frac{3}{4}$.

Figure 8.4 A number line partitioned into sixteenths illustrating that $\frac{3}{4}$ and $\frac{6}{8}$ and $\frac{12}{16}$ are all located at the same point on the number line, showing that they are different names for the same number

In Figure 8.5 Matthew was asked to identify two fractions that are located between $\frac{1}{3}$ and $\frac{3}{4}$. He responded by listing $\frac{2}{3}$ and $\frac{4}{6}$. What Mathew identified as two fractions are in fact two equivalent fractions representing the same number. When asked—are there any other fractions between $\frac{1}{3}$ and $\frac{3}{4}$?—Matthew (Figure 8.5) responded "Yes, you can just keep transforming $\frac{2}{3}$." While he did not show evidence of understanding the density of rational numbers (the focus of

Chapter 7), he did provide evidence of understanding that there is an infinite set of fractions equivalent to $\frac{2}{3}$.

Figure 8.5 Mathew's response—which shows an ability to find an infinite number of equivalent fractions equal to $\frac{2}{3}$

Yes, you can just keep transforming $\frac{2}{3}$ for ever

A Framework for the Development of Equivalence Concepts

"Conceptual understanding of equivalent fractions involves more than remembering a fact or applying a procedure" (Wong & Evans, 2007, p. 826): *that is, understanding equivalence as well as procedures for finding equivalent fractions, so important for the development of other concepts, should be built in a way that brings meaning to both. Researchers suggest developing the connections between the concept and procedure through interaction with models and manipulatives. Using the models and manipulatives helps to reveal patterns and relationships built on an awareness of the connections between the size and number of equal parts in a whole* (Behr et al., 1984; Behr & Post, 1992; Wong & Evans, 2007; Payne, 1976).

For example, Michelle (Figure 8.6) is ready for questions that capitalize on her models to help build her understanding of equivalence. In Michelle's model it is easy to see the relationship between the size and number of parts in the whole. Michelle may be ready to describe $\frac{2}{10} = \frac{1}{5}$ by saying that "fifths are twice as large as tenths" or "there are two-tenths in every fifth."

Figure 8.6 Michelle's response

There are some candies in a dish.

$\frac{2}{5}$ of the candies are chocolate.
$\frac{3}{10}$ of the candies are peppermint.

Are there more chocolate candies or more peppermint candies?

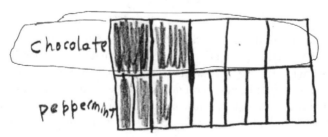

Questions for Michelle—such as "How many tenths are equal to $\frac{2}{5}$?", "How many fifths are equal to $\frac{8}{10}$?"—will help her to see the relationships between tenths and fifths. These questions are like standard questions in some drill exercises, perhaps in a form $\frac{2}{5} = \frac{n}{10}$ or $\frac{8}{10} = \frac{n}{5}$ but built on Michelle's developing understanding of equivalence in her models.

Taking this one step further, imagine that Michelle partitioned each tenth in half, resulting in her peppermint model partitioned into twentieths (Figure 8.7). Michelle could then consider how many twentieths are equal to $\frac{2}{5}$. The model in Figure 8.7 shows that $\frac{2}{5} = \frac{8}{20}$.

Figure 8.7 Michele's peppermint model partitioned into twentieths

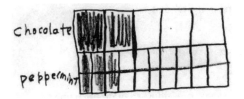

Michelle can begin to build a set of equivalent fractions $\left(\frac{2}{5} = \frac{4}{10} = \frac{8}{20}\right)$. This can be followed by additional partitioning to identify other fractions equivalent to $\frac{2}{5}$ (and $\frac{3}{5}, \frac{4}{5}$, and $\frac{5}{5}$). In this way Michelle can look for patterns in her models that eventually lead to an efficient procedure for finding equivalent fractions based on understandings that grew out of her models.

According to Van de Walle, the "goal is to help students see that by multiplying (or dividing) the top and bottom numbers by the same number, they will always get an equivalent fraction" (2004, p. 260).

The importance of building this understanding through patterns and relationships in models and not just teaching the procedure directly cannot be overemphasized. *Teachers have told us that because students are multiplying (or dividing) when they apply a procedure, they have a difficult time believing that equivalent fractions are really the same number* (VMP OGAP, personal communication, 2008).

Additionally, *some teachers indicated that they never understood the relationship between multiples, factors, partitioning of models, and equivalence that underpins finding common denominators and simplifying fractions. In their own words they were applying an algorithm without understanding* (VMP OGAP, personal communication, December, 2007).

Figure 8.8 uses Michelle's model from Figure 8.7 to illustrate the mathematical relationships that underpin the procedure. Using Michelle's model you will notice that there are four times as many equal parts in the whole partitioned into twentieths as there are in the whole partitioned into fifths. Likewise, there are four times as many equal parts in each fifth (and the parts are four times smaller). This multiplicative relationship forms the foundation of the procedure fully illustrated in Figure 8.8.

Figure 8.8 Based on Michelle's model in Figure 8.7 the following explanation describes the mathematical relationship that underpins the procedure of finding equivalent fractions by multiplying (or dividing) the numerator and the denominator by the same number

First . . .

Using Michelle's model in Figure 8.7 notice that when going from fifths to twentieths there are . . .

$\nearrow \times 4$ (four times as many equal parts in each fifth)
$$\frac{1}{5} = \frac{4}{20}$$
$\searrow \times 4$ (four times as many equal parts in the whole)

Because . . .

Multiplying the numerator and denominator each by 4 is the same as multiplying the fraction by 1 . . .
$$\frac{4}{4} = 1$$
Multiplying $\frac{1}{5} \times \frac{4}{4}(1)$ does not change the value of $\frac{1}{5}$, it just changes the name to $\frac{4}{20}$
$$\frac{1}{5} \times \frac{4}{4} = \frac{4}{20}$$

And . . .

Because *dividing by* 1 does not change the value of $\frac{4}{20}$,
to find out how many fifths is equivalent to four twentieths . . .
$$\frac{4}{20} \div \frac{4}{4} = \frac{1}{5}$$

Returning to Michelle's model . . .

Notice that the same region represents both one fifth and four twentieths; that is, they are equivalent fractions.

Over time, and with enough examples, students should see the multiplicative pattern consistent with all equivalent fractions that leads to the strategy of multiplying (or dividing) both the numerator and the denominator by the same number. To help students "believe" that equivalent fractions are different names for the same number, point out the obvious. In area models equivalent fractions are represented by the same region. On number lines, equivalent fractions are located at the same place, and, as you will see in the discussion related to sets of objects in Figures 8.9 to 8.11, equivalent fractions result in the same count of the objects in the set.

Figures 8.9 to 8.11 illustrate $\frac{1}{4} = \frac{2}{8} = \frac{4}{16}$ using a set of 32 apples partitioned into fourths (Figure 8.9), eighths (Figure 8.10), and sixteenths (Figure 8.11). Notice that the number of parts changes, not the number of apples that are circled; that is, $\frac{1}{4}$ of 32 apples is 8 apples, $\frac{2}{8}$ of 32 apples is 8 apples, and $\frac{4}{16}$ of 32 apples is 8 apples, because $\frac{1}{4} = \frac{2}{8} = \frac{4}{16}$. This is another way, in addition to a number line and area model, to show that the value of equivalent fractions is the same, and to illustrate again the mathematics essential to an understanding of equivalent fractions.

Figure 8.9 Circled here are $\frac{1}{4}$ of 32 apples. One fourth of 32 apples is 8 apples

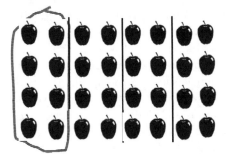

Figure 8.10 One column contains $\frac{1}{8}$ of the apples. Circled here are $2\left(\frac{1}{8}\right) = \frac{2}{8}$ of the apples. Two eighths of 32 apples is 8 apples

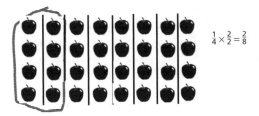

$$\frac{1}{4} \times \frac{2}{2} = \frac{2}{8}$$

Figure 8.11 One part contains $\frac{1}{16}$ of the apples. Circled here are $4\left(\frac{1}{16}\right) = \frac{4}{16}$ of the apples. Four sixteenths of 32 apples is 8 apples

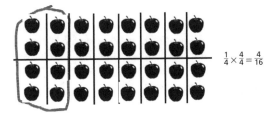

$$\frac{1}{4} \times \frac{4}{4} = \frac{4}{16}$$

Researchers have found that instruction which helps students to move flexibly between representations (spoken and written words (two-fifths), pictorial representations, manipulatives, contexts, and symbols) and within representations (e.g., $\frac{3}{4} = \frac{6}{8}$) will help students move toward equivalence reasoning that becomes free of the need to model (Post, Wachsmuth, Lesh, & Behr, 1985).

To illustrate what it means to move flexibly between representations, let's return to Michelle's model comparing $\frac{2}{5}$ and $\frac{3}{10}$ (Figure 8.6). If Michelle is to add $\frac{2}{5} + \frac{3}{10}$ (using symbols) she would need to apply her understanding derived from the model that $\frac{2}{5} = \frac{4}{10}$. This would allow her to add 3 tenths + 4 tenths = 7 tenths (represented in words). *Adding 3 tenths + 4 tenths is exactly the same as adding 3 dollars + 2 dollars = 5 dollars, or adding 3 hours + 2 hours = 5 hours* (Gross & Gross, 1999). Converting both fractions to tenths allows Michelle to add fractions represented by the same-sized pieces in her model, just like adding hours + hours. In this way intentional connections are made between the words, models, and symbols that bring meaning to each.

Transitioning students from models to a generalized symbolic representation begins as early as grades 3 and 4 in most mathematics programs. For example, as early as the second fraction lesson in grade 4, "Investigations Number, Data, and Space" (2007) (Investigations) students are asked to determine the relationships between thirds and sixths. First, they are asked to represent thirds and sixths with area models. Then they are asked to identify fractions that are equal to each other just as was suggested with Michelle's model. This type of connection continues throughout the fraction unit as students begin to compare fractions and then add and subtract fractions.

By the end of the fifth grade, students should have developed procedural fluency (built on modeling and reasoning) when solving problems involving the addition and subtraction with fractions (NCTM, 2006). This means that students need a firm understanding of equivalence as well as an efficient procedure to find equivalent fractions. Students who continue to rely solely on models to solve problems will be at a disadvantage (Figure 8.12).

Kyle's response (Figure 8.12) shows the limitation of being "stuck" with only a model. While all three fractions are equivalent, Trevor's inability to accurately partition sixteenths (understandably so) led him to conclude incorrectly that $\frac{12}{16}$ is not equivalent to $\frac{3}{4}$.

Figure 8.12 Kyle's response—which shows the limitation of only having modeling as a strategy for determining equivalence. In this case partitioning accurately to sixteenths is a limiting factor

Kim said that $\frac{3}{4}$ is equivalent to $\frac{6}{8}$ and to $\frac{12}{16}$. Is Kim correct?

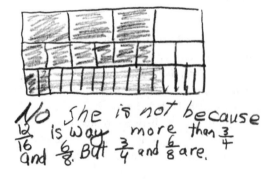

Conversely, Kieren (as cited in Huinker, 2002, pp. 73–4) found that *premature experience with formal procedures may lead to symbolic knowledge that is not based on understanding, or connected to the real world. This may impede students' number and operation sense.* As you have seen in the development of equivalence concepts and will read in Chapter 9, this is a delicate balance.

GO TO Chapter 9: Addition and Subtraction of Fractions for more information on developing student procedural and conceptual knowledge of addition and subtraction of fractions.

Chapter Summary

This chapter focused on the concept of equivalence and research related to developing an understanding of equivalence. Examples and discussions focused on:

- the meaning of equivalence;
- using models to develop conceptual understanding of equivalence; and
- the importance of transitioning students to an efficient and generalized strategy for finding equivalent fractions by the end of fifth grade for comparing fractions, ordering fractions, and for addition and subtraction of fractions.

Looking Back

1. What is the evidence in Emma's response in Figure 8.13 that demonstrates an understanding of equivalence in this situation?

Figure 8.13 Emma's response

Name one fraction that can be added to $\frac{1}{2}$ to get a sum of $\frac{7}{8}$.

2. Use models to address (a) and (b).
   (a) Illustrate that $\frac{2}{3}$, $\frac{4}{6}$, and $\frac{8}{12}$ are equivalent using area models, set models, and number lines.
   (b) Name one more fraction that is equivalent to $\frac{2}{3}$. Adapt one of your models in 2(a) to show that that fraction is equivalent to $\frac{2}{3}$.
3. Review Kenny's response (Figure 8.14) and then answer questions (a) and (b) that follow.
   (a) What is the evidence in Kenny's response in Figure 8.14 that demonstrates an understanding of equivalence in this situation?
   (b) How might a student select a model to use in solving this problem? Show how the model you select can help build an understanding of equivalence.
4. Chris accurately calculated the distance in the problem in Figure 8.15.
   (a) What is the evidence in Chris's response that he understands equivalence?
   (b) What concerns do you have about his solution?

Figure 8.14 Kenny's response

Tina ate $\frac{2}{3}$ of her candy and gave $\frac{1}{4}$ of her candy to her sister. She saved the rest of her candy. What is the fractional part of the candy that Tina saved?

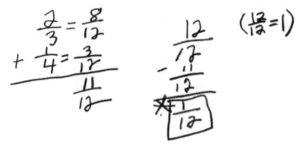

Figure 8.15 Chris's response

Billy drove $2\frac{1}{3}$ miles from his home to work. His car broke down $1\frac{4}{5}$ miles from work. How far was he from home?

$$2\frac{1}{3} - 1\frac{4}{5}$$

$$\frac{7\times5=35\quad 9\times3=27}{3\times5}$$

$$\frac{35}{15} - \frac{27}{15} = \boxed{\frac{8}{15}}$$

Instructional Link—Your Turn

Use the prompts in Table 8.1 to help you think about how your instruction and mathematics program provides students with the opportunity to develop understandings of equivalence.

Table 8.1 Instructional link strategies to support development of concepts related to equivalence.

Do you or does your program:	Yes/no
(1) use models to build understanding of equivalence?	
(2) transition from models to a generalized understanding of equivalence for comparing and ordering fractions?	
(3) transition from models to a generalized understanding of equivalence for adding and subtracting fractions?	

Identify any gaps between your instruction (including what your program offers) and what should be addressed, based on the research from this chapter.

Research Review—Equivalent Fractions

"Fraction order and equivalence ideas are fundamentally important concepts. They form the framework for understanding fractions and decimals as quantities that can be operated on in meaningful ways" (Post et al., 1993, p. 15).

Researchers say that *"students who really do not understand what a fraction means will have a hard time finding another fraction equivalent to it"* (Bezuk & Bieck, 1993, p. 129). In addition, some students *"have a continuing interference from their knowledge of whole numbers"* (Post, Behr, Lesh, & Wachsmuth, 1986).

"Conceptual understanding of equivalent fractions involves more than remembering a fact or applying a procedure" (Wong & Evans, 2007, p. 826); *that is, understanding equivalence as well as procedures for finding equivalent fractions, so important for the development of other concepts, should be built in a way that brings meaning to both.*

Researchers suggest developing the connections between the concept and procedure through interaction with models and manipulatives. Using models and manipulatives helps to reveal patterns and relationships built on an awareness of the connections between the size and number of equal parts in a whole (Behr, Wachsmuth, Post, & Lesh, 1984; Behr & Post, 1992; Wong & Evans, 2007; Payne, 1976).

According to Van de Walle, the "goal is to help students see that by multiplying (or dividing) the top and bottom numbers by the same number, they will always get an equivalent fraction" (2004, p. 260). *Teachers have told us that because students are multiplying (or dividing) when they apply the strategy, they have a difficult time believing that equivalent fractions are really the same number* (VMP OGAP, personal communication, 2008).

It has been at this junction that some OGAP teachers indicated that they never understood the relationship between multiples, factors, partitioning of models, and equivalence that underpins understanding of the application of equivalence to finding common denominators and simplifying fractions. In their own words they were applying an algorithm without understanding (VMP OGAP, personal communication, December, 2007).

Researchers have found that instruction which helps students move flexibly between representations (spoken and written words (two-fifths), pictorial representations, manipulatives, contexts, and symbols) and within representations (e.g., $\frac{3}{4} = \frac{6}{8}$) will help students to move toward equivalence reasoning that becomes free of the need to model (Post, Wachsmuth, Lesh, & Behr, 1985).

Adding 3 tenths + 4 tenths is exactly the same as adding 3 dollars + 2 dollars = 5 dollars, or adding 3 hours + 2 hours = 5 hours (Gross & Gross, 1999).

By the end of the fifth grade, students should have developed procedural fluency (built on modeling and reasoning) when solving problems involving the addition and subtraction with fractions (NCTM, 2006).

Kiernan (as cited in Huinker, 2002, pp. 73–4) *found that premature experience with formal procedures may lead to symbolic knowledge that is not based on understanding, or connected to the real world. This may impede students' number and operation sense.*

9

Addition and Subtraction
of Fractions

Big Ideas

- Procedural fluency and conceptual understanding work together to deepen student understanding of fraction addition and subtraction.
- Fraction addition and subtraction concepts build from, and are dependent upon, foundational part-to-whole, equivalence, and magnitude ideas.

Conceptual Understanding and Procedural Fluency

An important goal of fraction instruction is to ensure that students develop procedural fluency when adding and subtracting fractions. "*Procedural fluency refers to knowledge of procedures, knowledge of when and how to use them appropriately, and skill in performing them flexibly, accurately, and efficiently*" (National Research Council [NRC], 2001, p. 121).

It is important to understand, however, that procedural fluency alone is not sufficient to ensure proficiency with addition and subtraction of fractions. Procedural fluency works together with conceptual understanding, each contributing to a deeper understanding of the other.

> *Conceptual understanding refers to an integrated and functional grasp of mathematical ideas. Students with conceptual understanding know more than isolated facts and methods. They understand why a mathematical idea is important and the kinds of contexts in which it is useful.*

> (NRC, 2001, p. 118)

Figure 9.1 shows Kenny's strategy for solving a multistep fraction operation problem. Kenny finds and uses common denominators to add $\frac{2}{3} + \frac{1}{4}$, then uses his understanding that $\frac{12}{12} = 1$ to find the missing fractional part.

To determine Kenny's overall proficiency regarding adding and subtracting fractions, however, one would need to consider Kenny's understanding of addition and subtraction across contexts and with a variety of fractions (e.g., fractions with the same and different denominators, mixed numbers). That said, the evidence suggests that Kenny is on his way to becoming both procedurally fluent and conceptually sound with fraction addition and subtraction.

Figure 9.1 Kenny's response—the response shows an understanding of equivalent fractions

Tina ate $\frac{2}{3}$ of her candy and gave $\frac{1}{4}$ of her candy to her sister. She saved the rest of her candy. What is the fractional part of the candy that Tina saved?

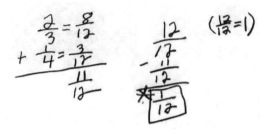

Fraction instruction that includes a thoughtful use of models and other reasoning strategies based on an understanding of the magnitude of fractions can lead to both procedural fluency and conceptual understanding of addition and subtraction of fractions. Researchers have found that students who can translate between various fraction representations *"are more likely to reason with fraction symbols as quantities and not as two whole numbers"* (Towsey, as cited in Huinker, 2002, p. 4) when solving problems.

The solutions in Figures 9.2 and 9.3 are examples of using an area model and a number line to solve addition and subtraction problems

Juan did not explicitly use common denominators to find the amount of candy that Tina saved, but evidence on his number line reveals a developing understanding of equivalence. For example, Juan partitioned his number line into thirds and twelfths. He used the partitioning to locate $\frac{1}{4}$ and $\frac{2}{3}$ on the number line, and then recognized that $\frac{1}{12}$ was left.

Figure 9.2 Felicia's response—Felicia used an area model to find the sum of $\frac{1}{4} + \frac{5}{8}$

Josh and Alison ordered one pizza. Josh ate $\frac{1}{4}$ of the pizza and Alison ate $\frac{5}{8}$ of the pizza. Did Josh and Allison eat the whole pizza?

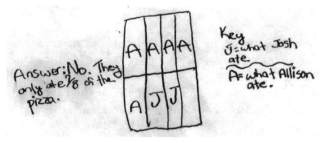

Figure 9.3 Juan's response—Juan used a number line in his solution to find the candy Tina saved

Tina ate $\frac{2}{3}$ of her candy and gave $\frac{1}{4}$ of her candy to her sister. She saved the rest of her candy. What is the fractional part of the candy that Tina saved?

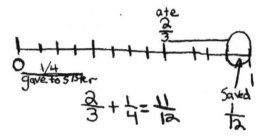

Both Felicia and Juan appear to be in a good position to move to a more efficient strategy that is grounded in conceptual understanding of partitioning, equivalence, and recognition of an additive situation.

Future instruction should include opportunities for these students to compare the information presented in a model with a calculation to ensure that they can:

- recognize the answer from a model;
- use an algorithm to solve fraction addition/subtraction problems;
- compare answers from calculations with answers from models;
- complete similar calculations without reference to a model.

Using a model is one way to solve the pizza problem in Figure 9.2. Another way to solve this problem is by reasoning involving an understanding of the magnitude of the fractions $\frac{1}{4}$ and $\frac{5}{8}$ particularly because an exact answer is not a requirement of the problem.

Sample reasoning strategies:

1. Five-eighths is three-eighths from the whole pizza. One-fourth is the same as two-eighths. That leaves one-eighth uneaten.
2. Five-eighths is one-eighth more than one-half $\left(\frac{4}{8}\right)$ of the pizza. One-fourth is the same as two-eighths. One-eighth (the amount of five-eighths over a half) + two-eighths do not equal one-half of the pizza. So, one-half of the pizza and less than one-half of a pizza is less than a whole pizza.
3. Five-eighths is three-eighths from a whole pizza. One-fourth of a pizza is less than three-eighths of the pizza. So $\frac{5}{8} + \frac{1}{4}$ is less than a whole pizza.

Many times instruction in adding and subtracting fractions focuses primarily on facility with algorithms and less on the conceptual underpinnings.

Premature experience with formal procedures may lead to symbolic knowledge that is not based on understanding, or connected to the real world. This may impede students' number and operation sense (Kiernan, as cited in Huinker, 2002).

Figures 9.4 and 9.5 contain pieces of student work that exemplify this point. You will notice that both students utilize an algorithm to solve the problems. The type of partial understandings and errors seen in these examples are typical of students who use algorithms without full understanding of the underlying concepts.

Felix (Figure 9.4) unnecessarily found a common denominator of 100 even though the fractions given in the problem already share a common denominator. In doing this, he may have made the problem unnecessarily harder.

Darcie (Figure 9.5) used an algorithm correctly to convert the given fractions to fractions with a common denominator and found the correct sum. However, her choice of 11 as being closest to $\frac{25}{24}$ provides evidence of a lack of understanding of the magnitude of the fractions in the problem or in the solution.

It is important to note that had Darcie been asked to compute only $\frac{3}{8} + \frac{2}{3}$, one might have been confident that she is well on her way to developing proficiency with fraction operations.

Figure 9.4 Felix's response—Felix found a common denominator even though each fraction given has 10 as a denominator

Jill walked her dog $\frac{3}{10}$ of a mile on Saturday and $\frac{4}{10}$ of a mile on Sunday. Is the total distance Jill walked her dog on Saturday and Sunday closest to $\frac{1}{2}$ of a mile or 1 whole mile?

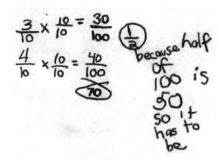

Figure 9.5 Darcie's response—Darcie calculated the correct sum but concluded that $\frac{25}{24}$ is closest to 11, not 1

$\frac{3}{8} + \frac{2}{3}$ is closest to

    A. 1

    B. 0

    C. 5

    D. 11

Of course, there is no way of knowing whether Darcie's solution is the result of premature focus on formal algorithms. However, research indicates that students can struggle with the use and understanding of formal algorithms *when their knowledge is dependent primarily on memory, rather than anchored with a deeper understanding of the foundational concepts* (Kieren, as cited in Huinker, 2002). Solutions like Darcie's serve as a reminder of the importance of focusing on understanding as students are developing efficient procedures.

Using Models, Partitioning, and Equivalence to Develop Fraction Addition and Subtraction Concepts

Even though instruction in addition and subtraction of fractions does not formally begin until the upper elementary grades, the roots of these operations begin in earlier grades as students partition models to represent fractional parts of a whole and compare fractions. Fraction addition and subtraction ideas are not isolated from fraction concepts explored in earlier grades, but rather, are a logical continuation of important part-to-whole and equivalence and magnitude ideas. *As students' understanding of fraction concepts grows, they often move from ordering and comparing fractions and finding equivalent fractions to adding and subtracting fractions* (VMP OGAP, personal communication, 2005).

The examples below, like those shown earlier, show evidence of developing understanding of part-to-whole, partitioning, and equivalence concepts.

In Figure 9.6 Holly states that "$\frac{3}{10}$ is just $\frac{1}{10}$ away from $\frac{2}{5}$." She is describing an additive relationship between $\frac{3}{10}$ and $\frac{2}{5}$. This relationship can be interpreted as $\frac{3}{10} + \frac{1}{10} = \frac{2}{5}$ or $\frac{2}{5} - \frac{1}{10} = \frac{3}{10}$. Holly's solution is an example of how fraction addition and subtraction concepts can logically develop from investigations of equivalence and magnitude.

It is important to note that without Holly's ability to model the given fractions, her attention to the size of the whole, and her understanding of partitioning, it would have been difficult for Holly to consider how much larger $\frac{2}{5}$ is than $\frac{3}{10}$. In this way, the part-to-whole skills and concepts she learned earlier can be built upon as she transitions into reasoning additively with fractions.

Figure 9.6 Holly's response—Holly created models and used equivalence to compare the relative size of $\frac{3}{10}$ and $\frac{2}{5}$

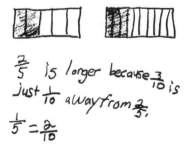

Figure 9.7 Patrick's response—Patrick used models to compare $\frac{2}{5}$ and $\frac{3}{10}$

There are some candies in a dish.

$\frac{2}{5}$ of the candies are chocolate.
$\frac{3}{10}$ of the candies are peppermint.

Are there more chocolate candies or more peppermint candies in the dish?

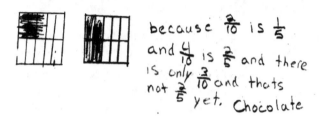

because $\frac{2}{10}$ is $\frac{1}{5}$ and $\frac{4}{10}$ is $\frac{2}{5}$ and there is only $\frac{3}{10}$ not $\frac{2}{5}$ $\frac{4}{10}$ and thats yet. Chocolate

Patrick's models in Figure 9.7 show that $\frac{2}{10} = \frac{1}{5}$ and $\frac{4}{10} = \frac{2}{5}$. This can be an important first step leading to an understanding of common denominators; a concept that is integral to developing an efficient algorithm to solve addition and subtraction problems. Although there is no evidence that Patrick contemplated addition or subtraction ideas, this work is ripe for those ideas to take root.

Teachers can facilitate this type of additive reasoning as students solve equivalence and magnitude problems by asking questions such as:

- How much greater or less is one fraction than another?
- How much would you have to add to or subtract from one fraction to equal the other?
- Create and describe a model that shows not only which fraction is greater, but also how much greater.
- Use your model to show equivalence.

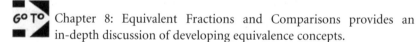 **GO TO** Chapter 8: Equivalent Fractions and Comparisons provides an in-depth discussion of developing equivalence concepts.

The Importance of Estimation When Adding and Subtracting Fractions

"The development of a quantitative notion, or an awareness of the "bigness" of fractions is very important" (Bezuk & Bieck, 1993, p. 127). Estimation plays a critical role in students' development of procedural fluency and conceptual understanding with fraction addition and subtraction. Procedural fluency includes the ability to recognize the most efficient way to solve a problem. A student with procedural fluency knows when and how to use a certain strategy depending on the problem and possesses the ability to use estimation to judge the reasonableness of an answer, as Lisa did in Figure 9.8.

Figure 9.8 Lisa's response—Lisa used her understanding of the "bigness" of the given fractions to answer the question correctly

Aunt Sally has a jar that holds one cup of liquid.
Her salad dressing recipe calls for $\frac{2}{3}$ cup of oil, $\frac{1}{8}$ cup of vinegar, and $\frac{1}{4}$ cup of juice.
Is the jar large enough to hold all the oil, vinegar, and juice?

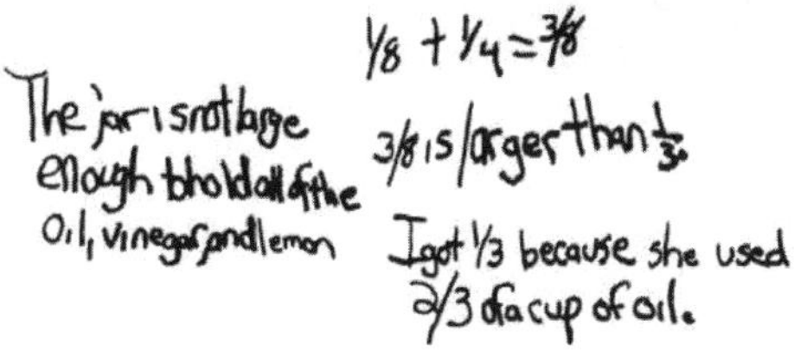

Lisa did not calculate an exact answer (although she did determine the sum of $\frac{1}{8} + \frac{1}{4}$) or create a concrete model to represent the situation. Instead, she used her understanding of the magnitudes of $\frac{1}{3}$, $\frac{3}{8}$, and $\frac{2}{3}$ to determine that the jar is not large enough to hold all three liquids.

Cody's solution shown in Figure 9.9 is correct and suggests facility with a fraction addition algorithm. However, it may not be the most efficient solution for this particular problem.

Cody's solution illustrates the notion that students often *do not apply their understanding of the magnitude (or meaning) of fractions when they operate with them* (NRC, 2001).

Using models to estimate a sum, as Oscar did in Figure 9.10, is a useful step along the continuum that leads to estimating sums in a more abstract manner. His experiences with modeling fractions can help Oscar to develop efficient estimation strategies, like the one Lisa (Figure 9.8) used, for determining the relative magnitude of fractions.

Figure 9.9 Cody's response—Cody calculated a specific sum even though an exact answer is not needed in this particular context

Aunt Sally has a jar that holds one cup of liquid.
Her salad dressing recipe calls for $\frac{2}{3}$ cup of oil, $\frac{1}{8}$ cup of vinegar, and $\frac{1}{4}$ cup of juice.
Is the jar large enough to hold all the oil, vinegar, and juice?

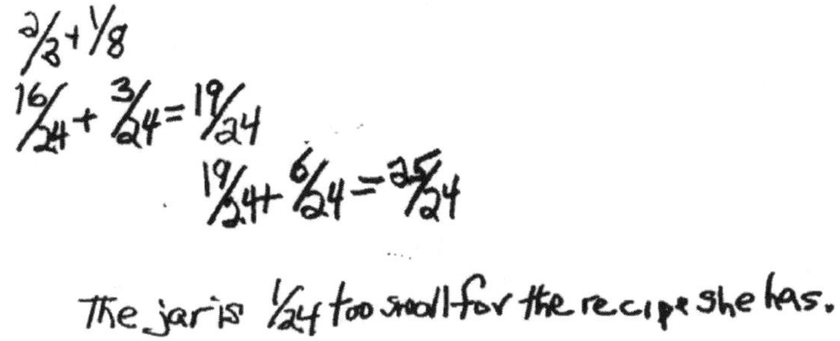

Figure 9.10 Oscar's response—Oscar did not calculate an exact answer. Instead, he used a model to show the relative size of $\frac{7}{8}$ and $\frac{1}{12}$

The sum of $\frac{1}{12} + \frac{7}{8}$ is closest to:

(a) 20
(b) 8
(c) $\frac{1}{2}$
(d) 1

An important instructional point is exemplified by Cody's, Oscar's, and Lisa's responses: "*students need facility with a variety of computational tools, and they need to know how to select the appropriate tool for a given situation*" (NRC, 2001, p. 122). Although each response includes a different strategy to answer the questions correctly, Lisa's strategy of reasoning about the quantities involved may be the most efficient given a context that does not require an exact numerical answer. Oscar's solution, using a physical model, might illustrate a developing understanding of estimating fraction sums. The evidence in Cody's response suggests that he is developing fluency with an algorithm when adding and subtracting fractions. Although this may not be the most efficient strategy for this problem, it may be quite efficient for a problem that requires an exact sum.

Adding and Subtracting Mixed Numbers

Up to this point, the chapter has focused on building conceptual understanding and procedural fluency with addition and subtraction of proper fractions. You have seen examples that illustrate how this knowledge can develop from equivalence and magnitude and part-to-whole concepts, and how the use of estimation and models can strengthen both conceptual and procedural knowledge of addition and subtraction. It is not surprising that these same ideas are fundamental to students' understanding of adding and subtracting mixed numbers and improper fractions.

We discussed earlier in the chapter how fraction addition and subtraction concepts can build from thoughtful exploration of equivalence and magnitude. Figures 9.11 and 9.12 illustrate this point.

Marcus may be in a good position to add $1\frac{1}{2}$ and $\frac{9}{8}$ because he has the sense of the magnitude of the two numbers relative to 1, and converted the improper fraction $\frac{9}{8}$ to $1\frac{1}{8}$. Based on the evidence in Marcus's work, one might expect that he would easily recognize the sum to be greater than 2.

Unlike Marcus's solution, the evidence in Gregory's work (Figure 9.12) leads one to believe that he does not understand the magnitude of the improper fraction $\frac{8}{7}$. Gregory ignores the fraction part of the mixed numbers and considers only the whole numbers.

Figure 9.11 Marcus's response—Marcus used the benchmark, 1, to compare the numbers

Susan ate $1\frac{1}{2}$ cupcakes and Billy ate $\frac{9}{8}$ cupcakes.
Who ate more cupcakes?

Susan ate more because $1\frac{1}{2}$ is bigger than $\frac{9}{8}$. I think $1\frac{1}{2}$ is bigger than $\frac{9}{8}$ because if you have $\frac{9}{8}$ it only goes over 1 by $\frac{1}{8}$. If you have $1\frac{1}{2}$ $\frac{1}{2}$ goes over one by $\frac{1}{2}$.

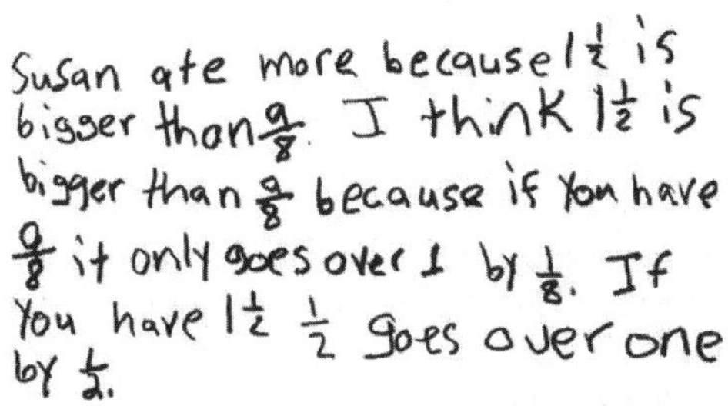

Figure 9.12 Gregory's response—Gregory appears to base his decision incorrectly on whether or not a whole number is explicit in the fraction notation given in the problem

Susan ate $1\frac{1}{6}$ cupcakes and Billy ate $\frac{8}{7}$ cupcakes.
Who ate more cupcakes?

Susan because she has 1 whole

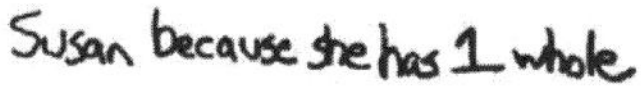

Gregory might benefit from opportunities to model fractions greater than 1 on number lines and with area models, and to generalize the relationships between mixed numbers and equivalent fractions. Marcus, on the other hand, appears to be ready to consider questions such as:

- How many cupcakes did Susan and Billy eat together?
- How many more cupcakes did Susan eat than Billy?

Tatsuoka completed an error analysis related to addition and subtraction of mixed numbers and improper fractions. Among the errors he identified is one represented in Renee's solution shown in Figure 9.13. *Students subtract the smaller fractional part of one mixed number from the larger fractional part of another regardless of the context* (Tatsuoka, 1984, pp. 9–12).

Another common error is identified by Tatsuoka (1984). *When subtracting mixed numbers students may borrow by reducing the whole number by one but incorrectly add that amount to the fraction part of the number (e.g., students add 10 to the denominator or ignore the value of the fraction part of the number).* This is exemplified in Shaun's response shown in Figure 9.14. He subtracted 1 from 2 but incorrectly added this amount to $\frac{1}{5}$. In performing the subtraction, he seemed to ignore the whole number part of the mixed number.

Figure 9.13 Renee's response—Renee ignored the whole number part of the mixed number and incorrectly subtracted the smaller fractional part, $\frac{1}{5}$, from the larger fraction, $\frac{3}{5}$

The distance from Billy's House to work is $2\frac{1}{5}$ miles.
His car broke down $\frac{3}{5}$ of a mile from work.
How far is Billy from his house?

$2\frac{1}{5} - \frac{3}{5}$ Answer° $2\frac{3}{5}$

$\frac{3}{5} - \frac{1}{5} = \frac{2}{5}$

Figure 9.14 Shaun's response—Shaun recognized that the problem could be solved using the operation $2\frac{1}{5} - \frac{3}{5}$. However, when borrowing, he seemed to ignore the fraction part of the mixed number, $2\frac{1}{5}$

The distance from Billy's House to work is $2\frac{1}{5}$ miles.
His car broke down $\frac{3}{5}$ of a mile from work.
How far is Billy from his house?

Gregory, Shaun, and Renee all made errors that led to an answer with an unreasonable magnitude given the problem (e.g., subtracting $\frac{3}{5}$ from $2\frac{1}{5}$ and obtaining an answer of $\frac{2}{5}$). For this reason all three students would benefit by representing their problems on a number line much as Lola did in Figure 9.15.

Notice that Lola's model suggests an understanding of the relative magnitude of the fractions presented in the problem. Lola is ready to use this conceptual understanding to develop a more generalized strategy for adding and subtracting mixed numbers.

Chapter Summary

This chapter focused on research related to adding and subtracting fractions. Through an examination of pertinent research and examples of student solutions, we demonstrated how:

- procedural fluency and conceptual understanding work together to deepen student understanding of fraction addition and subtraction;
- fraction addition and subtraction concepts build from, and are dependent upon, foundational part-to-whole and equivalence and magnitude ideas; and

Figure 9.15 Lola's response—Lola used a number line successfully to solve this problem involving mixed numbers

The distance from Billy's House to work is $2\frac{1}{5}$ miles.
His car broke down $\frac{3}{5}$ of a mile from work.
How far is Billy from his house?

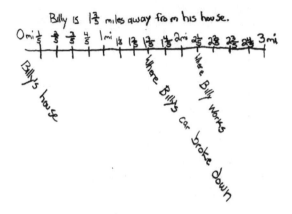

- estimation plays a critical role in students' development of procedural fluency and conceptual understanding as they relate to fraction addition and subtraction.

 Looking Back

1. Mrs. Grayson brought Kenny's work (first found in Figure 9.1 and again below in Figure 9.16) to a fifth-grade team meeting. She wondered if Kenny simply followed a procedure or if he understood the concepts upon which the algorithm is based.

    Help Mrs. Grayson by answering the following questions.

Figure 9.16 Kenny's response

Tina ate $\frac{2}{3}$ of her candy. She gave $\frac{1}{4}$ of her candy to her sister and saved the rest of her candy.

What is the fractional part of her candy that Tina saved?

$$\frac{2}{3} = \frac{8}{12}$$
$$+ \frac{1}{4} = \frac{3}{12}$$
$$\overline{\frac{11}{12}}$$

$$\frac{12}{12} \quad \left(\frac{12}{12} = 1\right)$$
$$- \frac{11}{12}$$
$$\overline{\frac{1}{12}}$$

    (a) Describe evidence(s) in Kenny's response that shows understanding of the context of the problem and related fraction concepts.

    (b) What questions might you ask Mrs. Grayson about her instruction to ensure that Kenny had a foundation for understanding?

    (c) If Mrs. Grayson wanted to be sure that Kenny understood the algorithm, what else could she ask him?

2. Mr. Benson brought Mathew's response (Figure 9.17) to the team meeting. He felt that this provides evidence that Mathew has a strong conceptualization when comparing $\frac{2}{5}$ to $\frac{3}{10}$ using both an area model and a number line.

    Answer the following questions.

    (a) What understandings are evidenced in Mathew's work?

    (b) What questions could be asked to build on understanding about equivalence and common denominators when comparing or adding and subtracting fractions? Explain how each question might help Mathew to move to a deeper understanding of equivalence and common denominators when comparing or adding and subtracting fractions.

3. Ms. Cunningham shared Kim's work (Figure 9.18) at the math team meeting. She is asking her teammates for advice on how to transition students to using models accurately to solve problems involving addition and subtraction.

    Help Ms. Cunningham by addressing the questions below:

    (a) What did Kim model correctly? What is the evidence?

    (b) Kim's model leads to an incorrect response. What errors did Kim make in her modeling? What is the evidence?

Figure 9.17 Mathew's response

There are some candies in a dish.

$\frac{2}{5}$ of the candies are chocolate.
$\frac{3}{10}$ of the candies are peppermint.

Are there more chocolate candies or more peppermint candies in the dish?

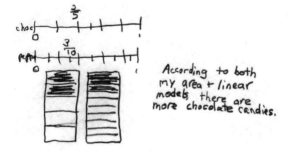

Figure 9.18 Kim's response

Tina ate $\frac{2}{3}$ of her candy. She gave $\frac{1}{4}$ of her candy to her sister and saved the rest of her candy. What is the fractional part of her candy that Tina saved?

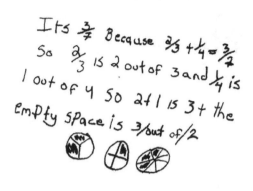

(c) What questions might you ask, or activities might you do, to help Kim understand how to use models to solve addition and subtraction problems?

4. Mr. Hill has been spending a lot of time working with his class on estimating sums and differences. He feels that the goal is for students to possess a "mental picture" of the magnitude of fractions, as Willy appears to have (Figure 9.19).

He wonders how he can help Oscar and Christine to move in this direction.

Consider the following questions.

(a) What did Christine (Figure 9.21) and Oscar (Figure 9.20) appear to understand? How do you know?

(b) What questions might you ask Christine so that she would not have to add the fractions before making the estimate? How would these questions help Christine?

(c) What questions (or activities) would you ask Oscar to help him move from using a model to a mental picture of the relative magnitude of the fractions being added? How would these help Oscar?

Figure 9.19 Willy's response

The sum of $\frac{1}{12} + \frac{7}{8}$ is closest to:

(a) 20
(b) 8
(c) $\frac{1}{2}$
(d) 1

I think 1 because $\frac{7}{8}$ is almost one $+ \frac{1}{12}$ is just going to be a little less than 1.

Figure 9.20 Oscar's response

The sum of $\frac{1}{12} + \frac{7}{8}$ is closest to:

(a) 20
(b) 8
(c) $\frac{1}{2}$
(d) 1

Figure 9.21 Christine's response

The sum of $\frac{1}{12} + \frac{7}{8}$ is closest to:

(a) 20
(b) 8
(c) $\frac{1}{2}$
(d) 1

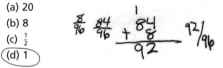

5. Ms. Horton is helping Emanuel (Figure 9.22) understand addition of proper fractions. He could draw models for most fractions and use models to add or subtract fractions with common denominators.

However, Emanuel (Figure 9.23) struggles with adding or subtracting fractions with unlike denominators. Help Ms. Horton by responding to the questions/prompts in (a)–(c).

Figure 9.22 Emanuel's response for adding fractions with common denominators

Solve $\frac{1}{8} + \frac{3}{8} =$

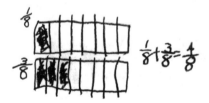

Figure 9.23 Emanuel's response for adding fractions with unlike denominators

Solve $\frac{1}{2} + \frac{3}{8} =$

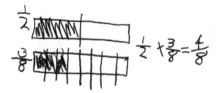

(a) What feature of the model in Figure 9.22 allowed Emmanuel successfully to add the two fractions that is not present in Figure 9.23?

(b) What would Emmanuel have to do to the model in Figure 9.23 to allow him effectively to use the same strategy for adding fractions as he used in Figure 9.22?

(c) Provide a sequence of addition/subtraction problems that would help build this understanding. Describe how the problems you propose can help to build an understanding of the meaning of common denominators.

Instructional Link—Your Turn

Use the questions in Table 9.1 to help you think about how your instruction and mathematics programs provide students the opportunity to develop understandings about addition and subtraction of fractions.

Identify any gaps between your instruction (including what your program offers) and what should be addressed based on the research from this chapter.

Consider instructional strategies that you could use to address the gaps you found.

Table 9.1 Instructional Link—strategies to support development of concepts related to addition and subtraction of fractions.

Do you or does your program provide opportunities for students to:	*Yes/no*
(1) use models to build understanding of addition and subtraction fraction concepts?	
(2) solve problems that involve estimating fraction sums and differences?	
(3) build upon other foundational skills such as modeling, partitioning, estimating, and equivalent fractions to develop both conceptual understanding and procedural fluency?	

Research Review—Addition and Subtraction of Fractions

"Procedural fluency refers to knowledge of procedures, knowledge of when and how to use them appropriately, and skill in performing them flexibly, accurately, and efficiently" (NRC, 2001, p. 121).

"Conceptual understanding refers to an integrated and functional grasp of mathematical ideas. Students with conceptual understanding know more than isolated facts and methods. They understand why a mathematical idea is important and the kinds of context in which it is useful" (NRC, 2001, p. 118).

Researchers have found that students who can translate between various fraction representations *"are more likely to reason with fraction symbols as quantities and not as two whole numbers"* (Towsey, cited in Huinker, 2002, p. 4) when solving problems.

Premature experience with formal procedures may lead to symbolic knowledge that is not based on understanding, or connected to the real world. This may impede students' number and operation sense (Kieren, cited in Huinker, 2002).

Research indicates that students can struggle with the use and understanding of formal algorithms *when their knowledge is dependent primarily on memory, rather than anchored with a deeper understanding of the foundational concepts* (Kieren, cited in Huinker, 2002).

As students' understanding of fraction concepts grows, they often move from ordering and comparing fractions and finding equivalent fractions to adding and subtracting fractions (VMP OGAP, personal communication, 2005).

"The development of a quantitative notion, or an awareness of the 'bigness' of fractions is very important" (Bezuk & Bieck, 1993, p. 127).

Students often *do not apply their understanding of the magnitude (or meaning) of fractions when they operate with them* (NRC, 2001).

"Students need facility with a variety of computational tools, and they need to know how to select the appropriate tool for a given situation" (NRC, 2001, p. 122).

Students subtract the smaller fractional part of one mixed number from the larger fractional part of another, regardless of the context (Tatsuoka, 1984).

When subtracting mixed numbers, students may borrow by reducing the whole number by one but incorrectly add that amount to the fraction part of the number (e.g., add 10 to the denominator or ignore the value of the fraction part of the number) (Tatsuoka, 1984).

10

Multiplication and Division
of Fractions

Big Ideas

- Multiplication and division of fractions are among the most complicated fraction concepts that elementary students encounter.
- Instructional opportunities that students encounter should include the meaning of multiplication and division in a range of situations and build procedural fluency with understanding.

Multiplication and Division of Fractions—Understanding the Concept

Fraction operations in general, and multiplication and division of fractions in particular, are consistently a source of confusion for students. *Research suggests that students often have a procedural knowledge of fraction operations but lack understanding of their vital underlying concepts* (Mack, cited in Yetkiner & Capraro, 2009). Why are multiplication and division of fractions so difficult for many students to understand? To begin to answer this question, let's explore aspects of the mathematics involved in these operations. In this chapter, all of the fractions used are positive fractions, and the word "fraction" always refers to a positive fraction.

Through the early elementary grades, students interact with multiplication and division problems involving the:

(a) identity property of multiplication—multiplication by 1 in which the product is equal to the other factor ($\underline{5} \times 1 = \underline{5}$);

(b) identity property of division—division by 1 resulting in a quotient that is equal to the dividend ($\underline{12} \div 1 = \underline{12}$);

(c) multiplication by a counting number greater than 1 resulting in a product that is greater than the other factor ($\underline{5} \times 4 = \underline{20}$);

(d) division by a counting number greater than 1 in which the quotient is less than the dividend ($\underline{12} \div 4 = \underline{3}$);

(e) zero property of multiplication—multiplication by zero equals zero;

(f) inverse relationship between multiplication and division;

(g) multiple interpretations of both multiplication (e.g., equal groups, area) and division (partitive and quotative—explained later in the chapter).

This set of understandings, while appropriate for work with whole numbers, does not consistently generalize to multiplication and division of fractions.

Many ideas that students develop about multiplication and division during their experiences with whole numbers will serve them well as they begin considering multiplication and division of fractions.

- The "identity property of multiplication" is true for both multiplication of fractions and whole numbers. That is, 1 times any number equals that number. For example, $8 \times 1 = 8$ and $\frac{1}{2} \times 1 = \frac{1}{2}$.
- The "zero property of multiplication" also holds true for multiplication of fractions as well as for whole numbers. This property states that zero times any number equals zero. Thus, $5 \times 0 = 0$ and $\frac{7}{8} \times 0 = 0$.
- Multiplication and division are inverse operations regardless of the numbers involved. This means that $2 \div \frac{1}{2} = 4$ and $4 \times \frac{1}{2} = 2$ for the same reason that $10 \div 2 = 5$ and $5 \times 2 = 10$. Often, students are asked to list members of the same multiplication and division "fact families." Just as $10 \div 2 = 5$ and $5 \times 2 = 10$ are in the same fact family, the equations $2 \div \frac{1}{2} = 4$ and $4 \times \frac{1}{2} = 2$ are in the same fact family.

As students begin operating with fractions, however, many have the wrong idea that multiplication always results in a larger number and division always results in a smaller number. Doug's response in Figure 10.1 is an example of a student who believes that multiplication must "make larger."

Figure 10.1 Doug's response—Doug states that the product has to be greater than the given factors

Stephanie and Paige are discussing the answer to $3\frac{2}{7} \times \frac{5}{9}$.
Stephanie said that the answer is more than $3\frac{2}{7}$.
Paige said the answer is less than $3\frac{2}{7}$.
Who is correct?

> if you multiplie any thing it has to
> be bigger than what you multiplie by.
> Stephanie is right.

To "undo" deeply held beliefs, like Doug's, about the impact of multiplication and division on a product (or quotient), instruction needs to focus on developing an understanding of why this belief does not always hold up for fractions. It is not enough just to tell Doug that it is not true. To explore this concept in more depth, we'll use the multiplication problem about halving a recipe that follows.

A recipe calls for $\frac{3}{4}$ of a cup of flour. How much flour is needed to make $\frac{1}{2}$ of the recipe?

Figure 10.2 uses an area model to represent the recipe problem. The area model and problem are used to illustrate two important points about multiplication and division of fractions:

1. Multiplication can result in a smaller product.
2. Multiplying by $\frac{1}{2}$ is the same as dividing by 2.

Figure 10.2 Model showing $\frac{1}{2} \times \frac{3}{4}$—One-half of $\frac{3}{4}$ of a cup of flour is represented in area model B

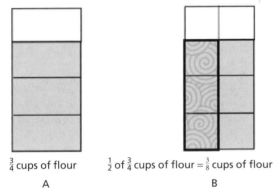

$\frac{3}{4}$ cups of flour $\frac{1}{2}$ of $\frac{3}{4}$ cups of flour $= \frac{3}{8}$ cups of flour

A   B

In Figure 10.2 the large rectangle represents a cup of flour. The area model A, represents $\frac{3}{4}$ of a cup of flour. The area model B, shows that $\frac{1}{2}$ of $\frac{3}{4}$ of a cup of flour equals $\frac{3}{8}$ of a cup of flour. It is easy to see in this case, particularly because of the understandable context, that a product in multiplication can be less than one of the factors. By extending this concept of multiplying a number by $\frac{1}{2}$ to multiplying a number by any fraction between 0 and 1, one sees that multiplication can make smaller. One is always finding a fractional part of a given number.

Model B also helps us to see that multiplying by $\frac{1}{2}$ is the same as dividing by 2; that is, when we consider $\frac{1}{2}$ of $\frac{3}{4}$ of a cup of flour, we either divide $\frac{3}{4}$ of a cup of flour by 2 or multiply $\frac{1}{2} \times \frac{3}{4}$ of a cup of flour. Both operations result in the correct answer, $\frac{3}{8}$ of a cup of flour. In either situation, the answer, $\frac{3}{8}$, is less than $\frac{3}{4}$, the original amount of flour.

One must not overgeneralize that anytime a fraction is involved in multiplication, the product will be less than the original number. For example, if one was preparing $2\frac{1}{2}$ times the recipe, the calculation would be $2\frac{1}{2} \times \frac{3}{4}$ cups of flour and the recipe would require more than $\frac{3}{4}$ cup of flour. See Figure 10.3.

Notice that the answer, $1\frac{7}{8}$ cups of flour, is greater than the original amount of flour, $\frac{3}{4}$ cups. This is because multiplication by a number greater than one, in this case $2\frac{1}{2}$, results in a product that is greater than the other factor.

Figure 10.3 Area models A and B each represent $\frac{3}{4}$ of a cup of flour. Area model C represents $\frac{1}{2}$ of $\frac{3}{4}$ of a cup of flour. The three models together represent $2\frac{1}{2} \times \frac{3}{4} = \frac{15}{8}$ or $1\frac{7}{8}$ cups of flour

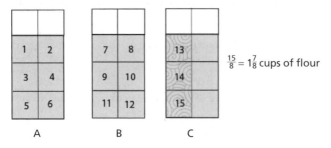

$\frac{15}{8} = 1\frac{7}{8}$ cups of flour

A   B   C

Impact of Division by a Fraction

As was described earlier in the chapter, students' experience with the division of whole numbers sometimes leads them to believe that the operation of division always "makes smaller." Glen's work in Figure 10.4 is an example of a student who is bringing a whole number notion of the impact of division to dividing by a fraction.

Figure 10.4 Glen's response—Glen chose the non-zero number less than $\frac{1}{2}$ and $\frac{1}{4}$ using his whole number understanding of the impact of division on a quotient

$\frac{1}{2} \div \frac{1}{4}$ is closest to?

(a) $\frac{1}{8}$

(b) 0

(c) 1

(d) 2

$\frac{1}{8}$ because 1,2 are larger Numbers and You Can't get 0 from dividing.

To investigate the impact of dividing by a fraction, we will use two problems. In each problem, we are dividing by a fraction that is less than 1. Here is the first problem.

Carly has $\frac{1}{2}$ pound of jelly beans. She filled bags with $\frac{1}{4}$ pound of jelly beans. How many bags did Carly fill?

The answer can be found by making the calculation $\frac{1}{2} \div \frac{1}{4}$. In this interpretation of division, one is asking, "How many $\frac{1}{4}$ pounds of jelly beans are in $\frac{1}{2}$ pound of jelly beans" or "How many $\frac{1}{4}$s are in $\frac{1}{2}$"? Figure 10.5 illustrates that there are 2 $\left(\frac{1}{4}s\right)$ in $\frac{1}{2}$, so $\frac{1}{2} \div \frac{1}{4} = 2$. Notice that the quotient, 2, is greater than the dividend, $\frac{1}{2}$. When a number is divided by a fraction less than 1, the quotient will be greater than the dividend.

Figure 10.5 Given the context in Carly's problem, the division expression $\frac{1}{2} \div \frac{1}{4}$ can be interpreted as: "How many one fourths are there in one half?" There are 2 ($\frac{1}{4}$s) in $\frac{1}{2}$, therefore Carly can fill two bags with jelly beans

$\frac{1}{2}$ lbs jellybeans $\div \frac{1}{4}$ lb per bag = 2 quarter lb bags of jellybeans

¼ lb of jellybeans	¼ lb of jellybeans

Now let's consider the second problem: $\frac{1}{4} \div \frac{1}{2}$.

The linear model in Figure 10.6 illustrates the impact of dividing a fraction by a fraction less than 1 when the divisor is greater than the dividend $\left(\frac{1}{4} \div \frac{1}{2}\right)$. Using the same interpretation of division as we used in the example in Figure 10.5, one can translate the expression $\frac{1}{4} \div \frac{1}{2}$ into "How much of $\frac{1}{2}$ is in $\frac{1}{4}$?"

In this case, as in the previous example, the quotient, $\frac{1}{2}$, is greater than the dividend, $\frac{1}{4}$.

It is worth repeating that for many elementary school students the idea that the answer to a division problem can be greater than the number being divided, or that multiplication can result in a smaller number is counterintuitive. Students will only come to this understanding after many opportunities to visualize the impact of dividing and multiplying by a fraction less than 1.

Figure 10.6 Linear model illustrating the calculation $\frac{1}{4} \div \frac{1}{2} = \frac{1}{2}$

How much of $\frac{1}{2}$ is in $\frac{1}{4}$?

Half of $\frac{1}{2}$ is in $\frac{1}{4}$

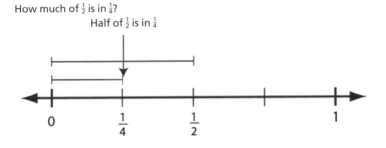

Justina's response in Figure 10.7 shows evidence of understanding the impact of multiplying by a fraction less than 1 on a product.

Figure 10.7 Justina's response—which shows evidence of understanding that multiplying by a number less than 1, like $\frac{4}{5}$, results in a smaller number

The product of $3 \times \frac{4}{5}$ is?

(a) greater than 3

(b) less than 3

less than 3 because in order for it to be more than 3, it would have to be multiplied by something greater than 1 but it is only multiplied by $\frac{4}{5}$.

Multiplication of Fractions: Understanding procedures

Students often learn to multiply fractions by simply multiplying the numerators and multiplying the denominators. Unfortunately, this type of instruction often leaves students with little conceptual understanding of this procedure. *Fraction computation can be taught as a series of rules; however, this focus on rote learning can result in artificial feelings of accomplishment* (Aksu, 1997).

The common algorithm for multiplying fractions involves multiplying the two numerators of the factors, and the two denominators to obtain a product.

$$\frac{1}{2} \times \frac{3}{4} = \frac{1 \times 3}{2 \times 4} = \frac{3}{8}$$

The calculation $\frac{1}{2} \times \frac{3}{4} = \frac{3}{8}$ is illustrated by the area model in Figure 10.8. Models such as these can be used to help students bring meaning to the procedure. There are two regions to consider in the model in Figure 10.8: (a) that representing the whole indicated by the large rectangle; and (b) that

Figure 10.8 Area model showing that $\frac{1}{2} \times \frac{3}{4} = \frac{3}{8}$

1 × 3 represents the number of eighths in the product $\frac{3}{8}$
$$\frac{1}{2} \times \frac{3}{4} = \frac{1 \times 3}{2 \times 4} = \frac{3}{8}$$
2 × 4 represents the number of eighths in the whole figure

representing $\frac{1}{2}$ of $\frac{3}{4}$ represented by the area with swirled shading. The product of the denominators $(4 \times 2 = 8)$ indicates the number of equal parts in the whole. The product of the numerators $(1 \times 3 = 3)$ indicates the number of eighths in the product $\frac{3}{8}$.

Division of Fractions: Understanding the Procedure

The two most widely used algorithms for division of fractions are the common denominator algorithm and the invert and multiply algorithm.

In order to understand how and why the common denominator algorithm works mathematically, we will look closely at the Bike Problem, the Fraction Problem, and the model in Figure 10.9 that follow.

Bike Problem

Chris rode her bike 10 miles. Kim rode her bike 2 miles. How many times as many miles did Chris ride than Kim?

Bike Answer

$$10 \text{ miles} \div 2 \text{ miles} = \frac{10 \text{ \sout{miles}}}{2 \text{ \sout{miles}}} = 5, \text{ or Chris rode 5 times as many miles}$$

as Kim.

The units in the Bike Problem are "miles." By dividing miles by miles we determined how many times more miles Chris rode than Kim. Notice that the units, miles, cancel each other out; that is, the answer is NOT in miles.

Fraction Problem

How many times greater is $\frac{1}{2}$ than $\frac{1}{3}$?

Fraction Answer

$$\frac{1}{2} \div \frac{1}{3} = \frac{3}{6} \div \frac{2}{6} = \frac{3 \text{ \sout{sixths}}}{2 \text{ \sout{sixths}}} = 3 \div 2 = \frac{3}{2} = 1\frac{1}{2}, \text{ or } \frac{1}{2} \text{ is } 1\frac{1}{2} \text{ times greater than } \frac{1}{3}.$$

The units in the Fraction Problem are "sixths." Dividing sixths by sixths results in how many times greater $\frac{3}{6}$ is than $\frac{2}{6}$. Notice that the units "sixths" cancel out; that is, the answer is NOT in sixths.

The area model in Figure 10.9 helps to visualize the unit "sixths." Notice that each "sixth" represents the same area. The model also helps us to understand that the answer will not be "sixths," but in the number of times greater that 3 sixths is than 2 sixths; $\frac{3 \text{ sixths}}{2 \text{ sixths}} = 1\frac{1}{2}$ times greater.

Figure 10.9 Illustrating the common denominator algorithm for the fraction problem $\frac{1}{2} \div \frac{1}{3}$

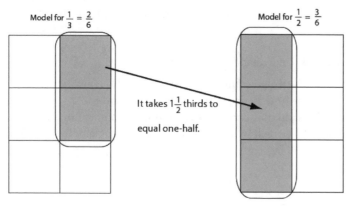

Tania (Figure 10.10) used the common denominator approach to dividing $4\frac{3}{8}$ yards of wire by $\frac{1}{2}$. In this example, Tania explicitly treated "eighths" as the unit for the calculation, and did not lose sight of the context of the problem that involved finding the number of decorations.

Figure 10.10 Tania's response—she calculated $35 \div 8$ to determine that Jim could make $8\frac{3}{4}$ decorations

Jim is making decorations. He has $4\frac{3}{8}$ yards of wire. Each decoration needs $\frac{1}{2}$ a yard of wire. How many decorations can Jim make?

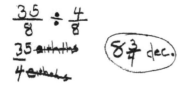

Earlier in this chapter (see Figure 10.2), we illustrated that multiplying a number by $\frac{1}{2}$ is the same as dividing the number by 2. This reciprocal relationship between $\frac{1}{2}$ and 2 can be used to understand the "invert and multiply" procedure for the division of fractions. Let us consider the following problem.

Toby hikes $\frac{3}{4}$ of a mile in $\frac{1}{3}$ of an hour.
How many miles does Toby walk in one hour?

Toby's rate (miles in one hour) can be found by making the calculation $\frac{3}{4}$ mile $\div \frac{1}{3}$ hr. Because division by a number is the same as multiplication by its reciprocal, we can also solve the problem using the expression, $\frac{3}{4}$ mile $\times \frac{3}{1}$ (or 3). Three fourths of a mile $\times$ 3 is the "invert and multiply strategy" for solving

$\frac{3}{4}$ mile $\div \frac{1}{3}$ hr. Figure 10.11 illustrates the multiplication of $\frac{3}{4}$ by 3 using a number line.

Figure 10.11 Linear model showing the relationship between hours and distance traveled. Toby walked $2\frac{1}{4}$ miles in one hour ($\frac{3}{4}$ miles $\times 3 = 2\frac{1}{4}$ miles in one hour)

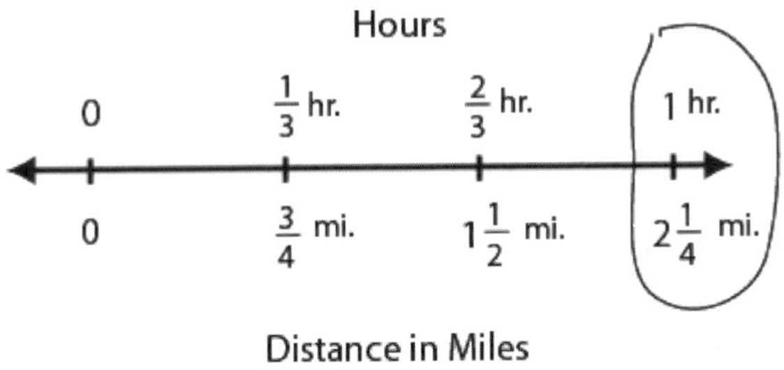

Hours

| 0 | $\frac{1}{3}$ hr. | $\frac{2}{3}$ hr. | 1 hr. |

| 0 | $\frac{3}{4}$ mi. | $1\frac{1}{2}$ mi. | $2\frac{1}{4}$ mi. |

Distance in Miles

As we observed with the example in Figures 10.2 and 10.11, dividing by the fraction, $\frac{a}{b}$ is the same as multiplying by its reciprocal, $\frac{b}{a}$. Students should be given plenty of opportunities to explore this relationship with a variety of fractions prior to being introduced to the invert and multiply algorithm.

Selma used the "invert and multiply" procedure to solve the problem in Figure 10.12.

Figure 10.12 Selma's response—Selma multiplied 6 pounds of candy by the reciprocal of $\frac{3}{4}$

Ashley bought 6 pounds of candy. She put the candy into bags that each hold $\frac{3}{4}$ of a pound of candy.
How many bags of candy did Ashley fill?

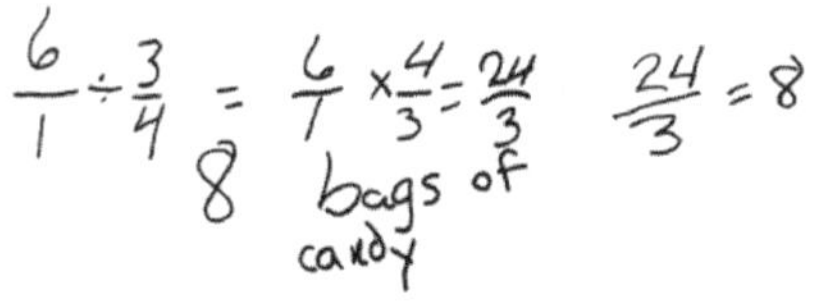

$$\frac{6}{1} \div \frac{3}{4} = \frac{6}{1} \times \frac{4}{3} = \frac{24}{3} \qquad \frac{24}{3} = 8$$

8 bags of candy

Anchoring Procedures in Understanding

As was discussed in Chapter 9, students may struggle with the use and under-standing of formal algorithms *when their knowledge is dependent primarily on memory, rather than anchored with a deeper understanding of the foundational concepts* (Kieren, cited in Huinker, 2002). *Some researchers even suggest that instruction focused on "rules" may present unintended consequences because rule-based instruction does not encourage students to think about the meaning of the operation. Mastery in the use of operations learned through rule-based instruction is quickly lost* (Aksu, 1997).

This research is particularly important considering the difficulty students have with multiplication and division of fractions. *Operations with fractions,* specifically division of fractions, are considered by some researchers to be the least understood topics in elementary school mathematics (Fendel, cited in Tirosh, 2000).

Abby's and Troy's incorrect application of algorithms in Figures 10.13 and 10.14 may be examples of premature reliance of procedures without conceptual understanding.

Figure 10.13 Abby's response—Abby divided 32 by the numerator and by the denominator of the fraction given

There are 32 students in the sixth-grade class. $\frac{5}{8}$ of the students are boys. How many boys are in the sixth-grade class?

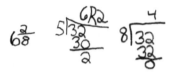

Figure 10.14 Troy's response—Troy calculated $2 \div \frac{3}{8}$ instead of $\frac{3}{8} \div 2$

Two friends equally share $\frac{3}{8}$ of a ball of yarn. How much of the yarn does each friend get?

$2 \div \frac{3}{8}$

$\frac{16}{8} \cdot \frac{3}{8}$

$\frac{16}{8} \times \frac{8}{3} = \frac{128}{24} = 5\frac{1}{3}$ division is just like fliping the second fraction and then multiplying

Figure 10.15 Kelyn's response—Kelyn interpreted the question as "How many $\frac{1}{4}$s are in four wholes?"

$4 \div \frac{1}{4}$ is closest to?

(a) 10
(b) 1
(c) 0
(d) 15

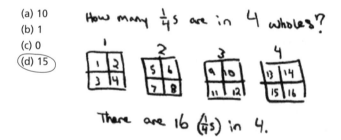

How many $\frac{1}{4}$s are in 4 wholes?

There are 16 ($\frac{1}{4}$s) in 4.

Contrast Abby's and Josh's response to Kelyn's solution in Figure 10.15. Kelyn's statement "How many $\frac{1}{4}$s are there in four wholes?" and her model provides evidence that Kelyn understands the problem $4 \div \frac{1}{4}$ and the impact of division by a fraction less than 1.

Proficiency with multiplication and division of fractions requires both a conceptual understanding of the meaning of the operations and the ability to use efficient strategies flexibly. Kelyn is ready to move to a more efficient procedure because she will bring meaning of division to the procedure.

Using Models to Build Understanding of Multiplication and Division

Considering the vital role that models can play in learning fraction concepts such as equivalence and magnitude, and addition and subtraction, it will come as no surprise that the same holds true for the importance of models in the teaching and learning of multiplication and division of fractions.

Researchers indicate that *teachers need knowledge of concrete models to help students' transition from multiplication by whole numbers to multiplication by fractions. Teachers must give closer consideration to division of fractions* (Taber, cited in Yetkiner & Capraro, 2009).

Figures 10.16 and 10.17 provide examples of students using models in their

Figure 10.16 Corey's response—Cory used a number line partitioned into $\frac{1}{4}$s to determine that eight bags can be filled

Ashley bought 6 pounds of candy. She put the candy into bags that each hold $\frac{3}{4}$ of a pound of candy. How many bags of candy did Ashley fill?

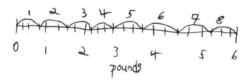

Figure 10.17 Jonathan's response—Jonathan partitioned $\frac{3}{8}$ of his model into two parts to represent $\frac{3}{8} \div 2$

Two friends equally share $\frac{3}{8}$ of a ball of yarn. How much of the ball of yarn did each friend get?

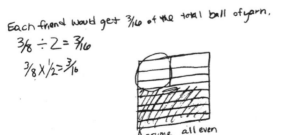

responses to fraction division problems. Although the models are different from one another, notice that each model accurately represents the context stated in the problem and leads to a correct response.

Mack (2001) suggests that *one cannot assume that students' prior experiences with models have prepared them to conceptualize more complex mathematical ideas such as multiplication and division of fractions and students may require guidance to reconsider these understandings.*

Tracy's and Gail's responses in Figures 10.18 and 10.19 may exemplify this point. Each drew and interpreted a model in a way that may have been appropriate when considering other fraction concepts. However, Tracy and Gail did not use the model to solve the division problems given as Kelyn did in Figure 10.15.

Figure 10.18 Tracy's response—Instead of determining how many $\frac{3}{4}$s are in 6 pounds, Tracy shaded $\frac{3}{4}$ of each bag

Ashley bought 6 pounds of candy. She put the candy into bags that each hold $\frac{3}{4}$ of a pound of candy. How many bags of candy did Ashley fill?

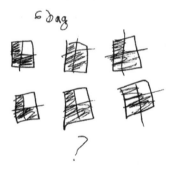

Figure 10.19 Gail's response—Gail shaded $\frac{1}{4}$ of four figures instead of finding how many $\frac{1}{4}$s are in 4

$4 \div \frac{1}{4}$ is closest to?

(a) 10
(b) 1
(c) 0
(d) 15

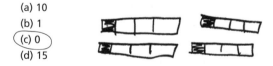

Even though Tracy's and Gail's models do not fully represent the problems they were solving, they can be used as starting points for instruction.

Both Tracy and Gail:

- represented the number of wholes consistent with their problems (although Tracy incorrectly labeled her wholes as bags not pounds);

- partitioned the wholes into fourths which could have allowed them each to solve their problems;
- shaded a fractional part of their wholes (e.g., Gail shaded $\frac{1}{4}$ of each of her wholes).

However, in both cases the students did not interpret the problems as division (e.g., Tracy found $\frac{3}{4}$ of each whole by shading $\frac{3}{4}$ of each figure. She did not find how many $\frac{3}{4}$s are in 6).

Clearly, instruction for Tracy and Gail needs to focus on the meaning of division and recognizing division in a problem situation involving fractions. However, one should capitalize on their ability to represent the whole and to partition into fourths.

Researchers suggest that *students should experience a variety of situations in which they need to recognize the appropriate fraction operation* (Huinker, 2002).

We have found that not only do some students struggle with recognizing when a context calls for a multiplicative solution, but students sometimes confuse the meaning and the procedures for multiplication and division of fractions (VMP OGAP, student work sample, 2009). Figures 10.20 and 10.21 below are examples of this confusion.

Figure 10.20 Alejandro's response—Alejandro modeled and described $\frac{1}{2} \times \frac{1}{4}$, not $\frac{1}{2} \div \frac{1}{4}$

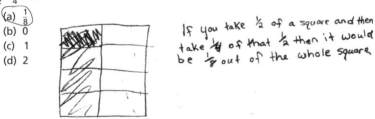

$\frac{1}{2} \div \frac{1}{4}$ is closest to:

(a) $\frac{1}{8}$
(b) 0
(c) 1
(d) 2

If you take ½ of a square and then take ¼ of that ½ then it would be ⅛ out of the whole square

Figure 10.21 Claudia's response—Claudia believes that the division sign indicates multiplication

$\frac{1}{2} \div \frac{1}{4}$ is closest to:

(a) $\frac{1}{8}$
(b) 0
(c) 1
(d) 2

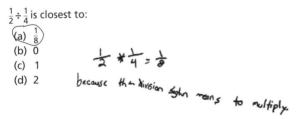

$\frac{1}{2} * \frac{1}{4} = \frac{1}{8}$

because the divsion sign means to multiply.

Both pieces of evidence imply that Alejandro and Claudia may not be clear about the difference between multiplying and dividing by a fraction. One wonders if either student has a conceptual knowledge of the meaning of division as it relates to fractions.

Multiplication and Division: Multiplicative and Additive Strategies

Procedural fluency in multiplication and division of fractions is an important goal of elementary and middle school mathematics. *It is important for computational procedures to be efficient, to be used accurately, and to result in correct answers* (NRC, 2001).

We will examine this idea of efficient procedures by analyzing Maya's and Dom's solutions to the candy problem shown in Figures 10.22 and 10.23.

Figure 10.22 Maya's response—Maya used repeated addition to determine that Ashley could fill eight bags

Ashley bought 6 pounds of candy. She put the candy into bags that each hold $\frac{3}{4}$ of a pound of candy. How many bags of candy did Ashley fill?

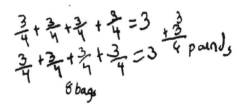

Figure 10.23 Dom's response—Dom used repeated subtraction to solve the problem

Ashley bought 6 pounds of candy. She put the candy into bags that each hold $\frac{3}{4}$ of a pound of candy. How many bags of candy did Ashley fill?

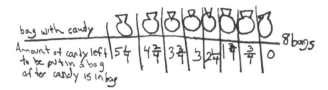

Maya used repeated addition until she reached 6 pounds.

Dom used repeated subtraction until he used up the 6 pounds of candy.

While Maya's and Dom's solutions are correct, the use of repeated addition and repeated subtraction will limit students' abilities to solve problems as the numbers increase in magnitude. Imagine Maya and Dom using these strategies if the question included 100 pounds of candy instead of 6. Viewed in this light, Tania's (Figure 10.10) and Selma's (Figure 10.12) procedures are more efficient to solve fraction division problems. Students who have not developed multiplicative strategies as have Tania and Selma to solve multiplication and division problems are at a significant disadvantage when faced with these more complex ideas and numbers.

Partitive and Quotative Division

A discussion of division of fractions would be incomplete without an examination of partitive and quotative division. *Partitive and quotative division are recognized as two different conceptual models for division* (Graeber & Tanenhaus, cited in Oksuz & Middleton, 2007).

In partitive division, also described as "fair sharing," the number of groups is known but the size of each group is unknown. Partitive division is often the first introduction that young students have to fraction concepts. In partitive division, the question being asked is: how much is in each group? An example is shown in Figure 10.24. (Note: the problem in Figure 10.11—Toby's hike—is another example of partitive division. In this case the question is "How much in 1 mile?")

Figure 10.24 Example of partitive division

Three friends share six pieces of candy.
How many pieces of candy does each friend receive?

In quotative division, the group size is known and the number of groups is unknown. The question asked in this situation is: how many groups are there in the whole? An example of quotative division is shown in Figure 10.25. (Note: the problem in Figure 10.5—Carly's jelly beans—is another example of quotative division.)

Figure 10.25 Example of quotative division

There are 6 yards of cloth.
Each pattern needs $\frac{2}{3}$ of a yard of cloth.
How many patterns can you make?

Partitive division can present unique challenges for students. In partitive division problems, students should *consider two questions; "How much is one share? and, What part of the unit is that share?"* (Lamon, 1999, p. 88).

In Figure 10.26, Cameron finds both the fraction of a pizza (one share) and the fraction of the pizzas (the unit).

In Figure 10.27, Brody does not correctly identify the size of one share and the fraction of the all the pizzas that each friend receives.

Brodi's model shows $\frac{1}{5}$ of one pizza, but is incomplete when considering the $3\frac{1}{2}$ pizzas. It is possible that he is wrestling with the conceptualization of a single pizza versus $3\frac{1}{2}$ pizzas.

Figure 10.26 Cameron's response—Cameron determined that each friend would get $\frac{7}{10}$ of a pizza and $\frac{1}{5}$ of the pizzas

Five friends equally share $3\frac{1}{2}$ pizzas.
(a) What fraction of a pizza does each friend get?

$3\frac{1}{2} = \frac{7}{2}$ or $\frac{35}{10}$ $\frac{1}{5}$ of $\frac{35}{10} = \frac{7}{10}$ So each person would get $\frac{7}{10}$ of a pizza.

(b) What fraction of all the pizzas does each friend get?

5 people So each person gets $\frac{1}{5}$ of the pizzas.

Figure 10.27 Brodi's response—Brodi concludes that each friend would receive $\frac{1}{5}$ of a pizza and $\frac{1}{5}$ of the pizzas

Five friends equally share $3\frac{1}{2}$ pizzas.
(a) What fraction of a pizza does each friend get?

 each get $\frac{1}{5}$

(b) What fraction of all the pizzas does each friend get?

They get $\frac{1}{5}$ of all the pizzas because if $3\frac{1}{2}$ is the whole then they each get $\frac{1}{5}$ of the whole

Quotative division can also pose challenges for students. More specifically, *students sometimes have a difficult time identifying the unit in quotative division problems* (Lamon, 1999). This difficulty can make it tricky for students to interpret a remainder.

In Figure 10.28 Cheney correctly used a number line to determine the number of full decorations. However, the evidence in Cheney's response indicates confusion with the remainder. The $\frac{2}{3}$ left is not $\frac{2}{3}$ of a yard, but $\frac{2}{3}$ of a decoration.

In contrast, Abigail's solution in Figure 10.29 suggests an understanding that $\frac{2}{3}$ refers to the fraction of a decoration that is left and that there is $\frac{1}{2}$ of a yard of wire remaining.

Figure 10.28 Cheney's response—Cheney used a number line to determine the number of full decorations, but misinterpreted the remainder

Jim is making decorations. He has $4\frac{1}{4}$ yards of wire. Each decoration needs $\frac{3}{4}$ of a yard of wire.

(a) How many full decorations can Jim make?

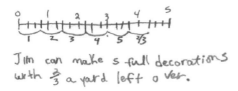

Jim can make 5 full decorations with $\frac{3}{4}$ a yard left over.

(b) Is the fraction left over a fraction of a decoration or a fraction of a yard of wire?

That fraction is 2/3 of a yard. The wire is in yards not decorations.

Figure 10.29 Abigail's response—Abigail uses a model to conclude that five full decorations can be made with $\frac{2}{3}$ of a decoration and $\frac{1}{2}$ of a yard of wire left over

Jim is making decorations. He has $4\frac{1}{4}$ yards of wire. Each decoration needs $\frac{3}{4}$ of a yard of wire.

(a) How many full decorations can Jim make?

(b) Is the fraction left over a fraction of a decoration or a fraction of a yard of wire?

there is $\frac{1}{2}$ of a yard left over or $\frac{2}{3}$ of a decoration because $\frac{1}{2} = \frac{2}{3}$ of $\frac{3}{4}$.

In the decoration problem students had to make meaning of a remainder in terms of the problem context. *Research has shown that "students' failure to solve division problems with remainders can be attributed, at least in part, to their failure to relate the computational results to the situation in the problem."* (Silver, Mukhopadhyay, & Gabriel, 1992, cited in Silver, Shapiro, & Deutsch, 1993)

From an instructional perspective it is important that the types of division problems that students solve be mixed up (Van de Walle, 2001). Students should encounter partitive problems, quotative problems, problems in which the remainders are fractions, problems in which remainders are whole numbers, problems with no remainders, and so on. Mixing problems in this way will help to ensure that students do not overgeneralize one way to think about division or one way to deal with remainders.

Chapter Summary

This chapter focused on research related to multiplying and dividing fractions. In particular we examined:

- the need to build procedural fluency through conceptual understanding, not through instruction focused solely on rote application of algorithms;
- the use of modeling and partitioning to help build multiplication and division concepts;
- the guidance and support that students need as they build understanding of complex concepts, such as multiplication and division of fractions, from ideas explored in earlier grades;
- the difficulties students encounter as they contemplate the impact of multiplication and division involving fractions on the magnitude of a product or quotient;
- the need for students to interact with a variety of situations and contexts that include both partitive and quotative division, and different kinds of remainder.

 Looking Back

1. Mr. Way gave his class a pre-assessment prior to the upcoming unit on multiplication and division of fractions. He is concerned about Claudia's response to the division problem in Figure 10.21. Help Mr. Way by answering the questions below.
   (a) What are some possible explanations for Claudia's apparent belief that "the division sign means to multiply?"
   (b) What are some questions, lessons, or activities that Mr. Way could use to help Claudia develop an understanding of the similarities and differences between multiplication of fractions and division of fractions?
2. The strategy shown in Figure 10.16 is representative of Corey's solutions for fraction division problems. Corey's teacher, Mrs. Rousseau, would like Corey to use her understanding of models to develop a more efficient algorithmic approach for division of fraction problems.

Examine Corey's response in Figure 10.16 and answer the questions that follow.

(a) Based on the evidence, what concepts related to division of fractions does Corey appear to understand?

(b) How could Mrs. Rousseau use the developing understandings you identified in Part (a) and her facility with models to help Corey develop an algorithmic approach for solving division of fractions problems?

3. Despite the fact that Ms. Altrui's class can use models effectively to solve equivalence, magnitude, addition and subtraction problems, the group is struggling with using models to solve more complex multiplication and division of fraction problems. Help Ms. Altrui by studying Tracy's solution in Figure 10.18 and answering the questions that follow.

(a) What context does Tracy's model appear to represent?

(b) How could Tracy's model be modified or reinterpreted to answer the question posed in the problem?

(c) Identify questions, activities, or lessons that Ms. Altrui could use to help her class extend their models of equivalence, magnitude, and addition and subtraction problems to include effective models for multiplication and division.

4. Although Alejandro can solve both multiplication and division of fraction problems, he tends to confuse the two operations. He often misinterprets problems requiring a division strategy and solves them using multiplication. Figure 10.20 above is an example of his confusion.

(a) How could Alejandro's model be altered or reinterpreted to answer the question, $\frac{1}{2} \div \frac{1}{4}$?

(b) How might you help Alejandro to conceptualize the similarities and differences between division by a fraction and multiplication by a fraction?

5. One of Selma's typical responses to division of fraction problems is shown in Figure 10.12. Mr. Latham, Selma's teacher, wants to be sure that Selma possesses the needed conceptual understanding of division of fractions to go along with her algorithmic knowledge.

(a) Based on the evidence in her response, what does Selma appear to understand about division of fractions?

(b) What questions might Mr. Latham ask Selma to help him to determine her conceptual understanding of division of fractions?

6. Mr. Alberti is preparing for an upcoming lesson on partitive division. As part of the lesson, he plans to use the question shown below. Mr. Alberti is contemplating a model that clearly shows the answers to both parts of the question.

Five friends equally share $3\frac{1}{2}$ pizzas.

    (a) What fraction of a pizza does each friend get?

    (b) What fraction of all the pizzas does each friend get?

  (a) Draw a model clearly showing that each friend gets $\frac{7}{10}$ of a pizza and $\frac{1}{5}$ of the pizzas.

  (b) Explain how you might connect the model you drew for Part (a) to the mathematical calculations $3\frac{1}{2} \div 5$ and $1 \div 5$.

7. Cheney's solution to a quotative division problem is shown in Figure 10.28.

  (a) What do the numbers on the top of Cheney's number line represent?

  (b) What do the numbers on the bottom of Cheney's number line represent?

  (c) What instructional strategies might you use to help Cheney understand that the fraction $\frac{2}{3}$ represents $\frac{2}{3}$ of a decoration, not $\frac{2}{3}$ of a yard?

  (d) How might you show how this problem results in both $\frac{1}{2}$ a yard and $\frac{2}{3}$ of a decoration left over?

Instructional Link—Your Turn

Table 10.1 Instructional Link—strategies to support development of concepts related to multiplication and division of fractions.

Do you or does your program . . .	Yes/no
(1) provide opportunities for students to develop conceptual understanding of multiplication and division of fractions before introducing formal algorithms?	
(2) provide opportunities for students to create and interact with concrete models to help their transition from multiplication and division by whole numbers to multiplication and division by fractions?	
(3) build on students' prior experiences with models?	
(4) provide opportunities for students to consider why both multiplication and division can "make smaller" and "make larger?"	
(5) provide a variety of situations in which students are asked to recognize the appropriate fraction operation?	
(6) provide students with contexts to reason multiplicatively without relying on additive strategies?	
(7) provide opportunities at the appropriate time for students to develop computational procedures that are efficient, accurate, and result in correct answers?	
(8) provide opportunities for students to translate mathematical ideas between real-world situations, manipulatives, pictures, spoken symbols, and written symbols?	

(9) provide opportunities for students to solve fraction problems involving both partitive and quotative division?

What gaps in your instruction or mathematics program did you identify? How might you address these gaps?

Research Review—Multiplication and Division of Fractions

Research suggests that students often have a procedural knowledge of fraction operations but lack understanding of their vital underlying concepts (Mack, cited in Yetkiner & Capraro, 2009).

Students may struggle with the use and understanding of formal algorithms *when their knowledge is dependent primarily on memory, rather than anchored with a deeper understanding of the foundational concepts* (Kieren, cited in Huinker, 2002).

Some researchers even suggest that instruction focused on "rules" may present unintended consequences because rule-based instruction does not encourage students to think about the meaning of the operation. Mastery in the use of operations learned through rules-based instruction is quickly lost (Aksu, 1997).

Operations with fractions, specifically division of fractions, are considered by some researchers to be the least understood topics in elementary school mathematics (Fendel, cited in Tirosh, 2000).

Researchers indicate that *teachers need knowledge of concrete models to help students' transition from multiplication by whole numbers to multiplication by fractions. Teachers must give closer consideration to division of fractions* (Taber, cited in Yetkiner & Capraro, 2009).

Mack (2001) suggests that *one cannot assume that students' prior experiences with models have prepared them to conceptualize more complex mathematical ideas such as multiplication and division of fractions and students may require guidance to reconsider these understandings.*

Researchers suggest that *students should experience a variety of situations in which they need to recognize the appropriate fraction operation* (Huinker, 2002).

We have found that not only do some students struggle with recognizing when a context calls for a multiplicative solution, but students sometimes confuse the meaning and the procedures for multiplication and division of fractions (VMP OGAP, student work sample, 2009). *It is important for computational procedures to be efficient, to be used accurately, and to result in correct answers* (NRC, 2001).

Partitive and quotative division are recognized as two different conceptual models for division (Graeber & Tanenhaus, cited in Oksuz & Middleton, 2007).

In partitive division problems, students should *consider two questions: "How much is one share? and, What part of the unit is that share?"* (Lamon, 1999, p. 88).

Students sometimes have a difficult time identifying the unit in quotative division problems (Lamon, 1999). *Research has shown that "students' failure to solve division problems with remainders can be attributed, at least in part, to their failure to relate the computational results to the situation in the problem"* (Silver, Mukhopadhyay, & Gabriel, 1992, cited in Silver, Shapiro, & Deutsch, 1993).

From an instructional perspective it is important that the types of division problem that students solve should be mixed up (Van de Walle, 2001).

References

Aksu, M. (1997). Student performance in dealing with fractions (electronic version). *Journal of Educational Research, 90(6)*, 375–80.

Behr, M. & Post, T. (1992). Teaching rational number and decimal concepts. In T. Post (Ed.), *Teaching mathematics in grades K-8: Research-based methods* (2nd ed.) (pp. 201–48). Boston: Allyn and Bacon.

Behr, M., Wachsmuth, I., Post T., & Lesh R. (1984). Order and equivalence of rational numbers: A clinical teaching experiment. *Journal for Research in Mathematics Education, 15(5)*, 323–341.

Bezuk, N. S. & Bieck, M. (1993). Current research on rational numbers and common fractions: Summary and implications for teachers. In D. T. Owens (Ed.), *Research ideas for the classroom: Middle grade mathematics* (pp. 118–36). New York: Macmillan.

Bright, G., Behr, M., Post, T., & Wachsmuth, I. (1988). Identifying fractions on number lines. *Journal for Research in Mathematics Education, 19(3)*, 215–232.

Clements, D. (1999). Concrete manipulatives, Concrete ideas. *Contemporary Issues in Early Childhood, 1(1)*, 45–60.

Fosnot, C. & Dolk, M. (2002). *Young mathematicians at work: Constructing fractions, decimals, and percents.* Portsmouth, NH: Heinemann.

Gross, H. & Gross, K. (1999). *Mathematics as a second language.* Unpublished manuscript, University of Vermont.

Huinker, D. (2002). Examining dimensions of fraction operation sense. In B. Litwiller (Ed.), *Making sense of fractions, ratios, and proportions, 2002 yearbook* (pp. 72–78). Reston, VA: NCTM.

Kribs-Zaleta, C. (2006, November 9) *Invented strategies for division of fractions.* Paper presented at the annual meeting of the North American Chapter of the International Group for the Psychology of Mathematics Education. Retrieved January 15, 2009, from http://www. allacademic.com/meta/p115409_index.html

Lamon, S. (1999). *Teaching fractions and ratios for understanding: Essential content and instructional strategies for teachers.* Mahwah, NJ: Lawrence Erlbaum Associates.

Lesh, R., Landau, M., & Hamilton, E. (1983). Conceptual models in applied mathematical problem solving research. In R. Lesh & M. Landau (Eds.), *Acquisition of mathematics concepts & processes* (pp. 263–343). New York: Academic Press.

Mack, N. (2001). Building on informal knowledge through instruction in a complex content domain: Partitioning, units, and understanding multiplication of fractions. *Journal for Research in Mathematics Education, 32(3)*, 267–295.

Mitchell, A. & Horne, M. (2008). Fraction number line tasks and the additivity concept of length measurement. In M. Goos, R. Brown, & K. Makar (Eds.), *Navigating currents and charting directions: Proceedings of the 31st annual conference of the Mathematics Education Research Group of Australasia* (pp. 353–360). Brisbane, Queensland, Australia: Mathematics Education Research Group of Australasia.

National Council of Teachers of Mathematics (2006). *Curriculum focal points for prekindergarten through grade 8: A quest for coherence.* Reston, VA: NCTM.

National Research Council (2001). *Adding it up: Helping children learn mathematics.* J. Kilpatrick, J. Swafford, & B. Findell (Eds.), Washington, DC: National Academies Press.

Oksuz, C. & Middleton, J. (2007). Middle school children's understanding of algebraic fractions as quotients. *International Online Journal of Science and Mathematics Education, 7*, 1–14. Retrieved March 28, 2009, from http://www.upd.edu.ph/~ismed/online/articles/middle/Vol7_Middle.pdf

Orton, R., Post, T., Behr, M., Cramer, K., Harel, G., & Lesh, R. (1995). Logical and psychological aspects of rational number pedagogical reasoning. *Hiroshima Journal of Mathematics Education, 3*, 63–75.

Payne, J. N. (1976). Review of research on fractions. In R. Lesh & D. Bradbard (Eds.), *Number and measurement: Papers from a research workshop* (pp. 145–187). Columbus, Ohio: Information Reference Center.

Petitto, A. (1990). Development of numberline and measurement concepts. *Cognition and Instruction, 7(1)*, 55–78.

Post, T. (1981). The role of manipulative materials in the learning of mathematical concepts. In M. Lindquist, *Selected Issues in Mathematics Education* (pp. 109–131). Berkeley, CA: McCutchan.

Post, T., Behr, M., Lesh, R., & Wachsmuth, I. (1986, Spring). Selected results from the rational number project. In Proceedings of The Ninth Psychology of Mathematics Education Conference, The Netherlands (pp. 342–351). International Group for the Psychology of Mathematics Education, ANTWERP The Netherlands. This paper was reprinted in *The Math Times Journal—Official Journal of the Minnesota Council of Teachers of Mathematics, 1*, (1).

Post, T., Cramer, K., Behr, M., Lesh, R., & Harel, G. (1993). Curriculum implications of research on the learning, teaching, and assessing of rational number concepts. In T. Carpenter, E. Fennema & T. Romberg (Eds.), *Rational numbers: An integration of research* (pp. 107–130). Mahwah, NJ: Lawrence Erlbaum and Associates.

Post, T. & Reys, R. E. (1979). Abstraction generalization and design of mathematical experiences for children. In K. Fuson & W. Geeslin (Eds.), *Models form mathematics learning* (pp. 117–139). Columbus, OH: ERIC/SMEAC.

Post T., Wachsmuth I., Lesh R., & Behr, M. (1985). Order and equivalence of rational number: A cognitive analysis. *Journal for Research in Mathematics Education, 16(1)*, 18–36.

Pothier, Y., & Sawada, D. (1983). Partitioning: The emergence of rational number ideas in young children. *Journal for Research in Mathematics Education, 14(5)*, 307–317.

Saxe, G. B., Shaughnessey, M., Shannon, A., Langer-Osuna, J., Chinn, R., & Gearhart, M. (2007). Learning about fractions as points on a number line. In W.G. Martin, M.E. Strutchens, & P. C. Eliot (Eds.), *The learning of mathematics 2007 yearbook* (pp. 221–236). Reston, VA: National Council of Teachers of Mathematics.

Silver, E., Shapiro, L., and Deutsch, A. (March, 1993). Sense making and the solution of division problems involving remainders: An examination of middle school students' solution processes and their interpretations of solutions. *Journal for Research in Mathematics Education, 24*, (2), 117–135 (Reston, VA: National Council of Teachers of Mathematics).

Tatsuoka, K., Ed. (1984). *Analysis errors in fraction addition and subtraction problems: Final report*, Urbana, IL: University of Illinois. (ERIC Document Reproduction Service No. ED257665.)

Tirosh, D. (2000). Enhancing prospective teachers' knowledge of children's conceptions: The case of division of fractions. *Journal for Research in Mathematics Education, 31(1)*, 5–25.

Tirosh, D., Fischbein, E., Graeber, A., & Wilson, J. (1998). *Prospective elementary teachers' conception of rational number*, Retrieved May 10, 2009 from The University of Georgia, Mathematics Education website: http://jwilson.coe.uga.edu/texts.folder/tirosh/pros.el.tchrs.html.

Van de Walle, J. (2001). *Elementary and middle school mathematics: Teaching developmentally* (5th ed.). Boston: Pearson.

Vermont Mathematics Partnership Ongoing Assessment (2005). Exploratory study, Unpublished raw data.

Vermont Mathematics Partnership Ongoing Assessment (2006–2007). Fraction scale-up, Unpublished raw data.

Vermont Mathematics Partnership Ongoing Assessment Materials and Resources (2005–2009). *OGAP questions and student work samples.* Unpublished manuscript.

Viadero, D. (2007, April 24). Studies find that use of learning toys can backfire. *Education Week*, 12–13.

Wong, M., & Evans, D. (2007). Students' conceptual understanding of equivalent fractions. In J. Watson & K. Beswick (Eds.), *Mathematics: Essential research, essential practice*, Proceedings of the 30th Annual Conference of the Mathematics Education, Research Group of Australasia, vol. 2, pp. 824–833.

Yetkiner, Z. E. & Capraro, M. M. (2009). *Research summary: Teaching fractions in middle grades.* Retrieved April 3, 2009 from http://www.nmsa.org/Research/ResearchSummaries/TeachingFractions/tabid/1866/ Default.aspx.

Index